MONSTERS

ALSO BY ED REGIS

Regenesis (with George M. Church)

What Is Life?

The Info Mesa

The Biology of Doom

Virus Ground Zero

Nano

Great Mambo Chicken and the Transhuman Condition

Who Got Einstein's Office?

MONSTERS

THE *HINDENBURG* DISASTER
AND THE BIRTH OF
PATHOLOGICAL TECHNOLOGY

ED REGIS

BASIC BOOKS
A Member of the Perseus Books Group
New York

Published by Basic Books,
A Member of the Perseus Books Group

Designed by Trish Wilkinson
Set in 11.5 point Fairfield LT Std

Library of Congress Cataloging-in-Publication Data

Regis, Edward, 1944–
    Monsters : the Hindenburg disaster and the birth of pathological technology /
Ed Regis.
        pages   cm
    Includes bibliographical references and index.
    ISBN 978-0-465-06594-3 (hardcover)—ISBN 978-0-465-06160-0 (e-book)
1. Hindenburg (Airship) 2. Aircraft accidents. 3. Technological
innovations—Social aspects. 4. Technology—History. 5. Technology—Risk
assessment. I. Title.
TL659.H5R44 2015
303.48'30904—dc23   2015022692

10 9 8 7 6 5 4 3 2 1

The *Hindenburg* over Manhattan, May 6, 1937, three hours before its destruction.

To Pam

CONTENTS

Contents

PART III

ENDINGS

THE PATHOLOGY

The destruction of the airship *Hindenburg* on May 6, 1937, at Lakehurst, New Jersey, was one of the most visually shocking events of the midtwentieth century: it was the first time that a large-scale disaster was recorded on film, then coupled to a live, first-person account of the unfolding drama, and the result presented to the masses on larger-than-life motion picture screens. The combined image and soundtrack describing a scene of wild fury—untamed, uncontrolled, and unstoppable—instantly became an iconic, defining symbol of the era.

But that cataclysm, overlaid with portents of doom, also represented something else that has hitherto been little noticed, although it was almost as obvious, and just as striking, in its own way. The *Hindenburg* was an example of superior engineering, an object of advanced technology. Despite those attributes, it became a smoldering wreck in a bit more than thirty seconds. That should not have come as a surprise to its builders or operators, or for that matter even to its passengers. For one thing, the physical makeup of the craft virtually foretold and predetermined its fate: the *Hindenburg* was an immense vessel filled with more than seven million cubic feet of hydrogen gas, a highly flammable, indeed explosive, substance. For another, the craft's fiery ending had been preceded by a number of prequels, many of them equally if not more spectacular, and some of them even more deadly, although none of them had been captured on film. In fact, prior to the *Hindenburg* disaster, a total of twenty-six hydrogen airships had been destroyed by fire due to accidental causes, sometimes killing every last person aboard. Yet hydrogen-inflated zeppelins continued to be built and routinely flown in commercial passenger service. How did it come about that so much time, money, and labor were spent designing, building, maintaining, and operating a craft that was continuously risking the lives of its passengers and constantly flirting with death?

And it was not only German zeppelins that ended their lives as blackened and charred metal husks. The British built and operated their own hydrogen-filled airships, some of which broke up in midair; one of them killed forty-four of its passengers, even more than were lost in the *Hindenburg*. In 1930 the immense R 101 set off from the Royal Airship Works at Cardington on its maiden long-distance flight, which was to India. Aboard was the British air minister, Lord Thomson, who had said of the craft: "She is safe as a house—except for the millionth chance." Eight hours after takeoff the R 101 crashed into a hillside in France, where it exploded and burned, killing forty-eight of those aboard, including Lord Thomson.

Plainly, there was something deeply wrong with an oversize technological artifact that regularly put millions of cubic feet of explosive gas into close proximity with live and innocent human beings. The *Hindenburg* was a prototypical example of a *pathological technology*, one whose obvious and sizable risks were ignored, discounted, minimized, and swept under the rug by the influence of what amounted to an overriding, all-consuming, and almost irresistible emotional infatuation. That mental aberration produced a cognitive blindness to the craft's systemic defects. The *Hindenburg* and other zeppelins were built and flown because the bigger these leviathans got the more they acquired a spellbinding, mesmerizing, hypnotizing, practically immobilizing sway over human minds and emotions—not only those of the public and paying passengers, but also of their designers, builders, and crew members.

Indeed, even *after* the *Hindenburg* disaster, which was the most hair-raising public calamity up to that point, the Germans did not abandon their cherished invention. Instead, the Zeppelin Transport Company (owners of the *Hindenburg*) inflated with explosive hydrogen a *second* airship of equal size, the *Graf Zeppelin II,* and sent it aloft on thirty flights, many of them under the command of Captain Albert Sammt, who himself had been burned and scarred for life in the *Hindenburg* inferno. This, too, was pathological. In fact, the *Graf Zeppelin II* was *never* freely and voluntarily abandoned by its operators; instead, it was scrapped in 1940 only because Reich air minister Hermann Goering had decided that the ship's metal framework and components could be more fittingly used in bombers.

Never has a technology been so soundly, thoroughly, and utterly discredited as the hydrogen airship. Nothing whatsoever arose from its ashes. The craft was a complete and final dead end. But any technology that is so totally invalidated by a forty-year-long course of adverse events gives rise to a number of questions: How did it ever get started in the first place? How did it reach the ultra-high-risk degree of development that it did? And why did the technology persist for so long even in the face of its obvious and considerable flaws? The hydrogen airship also raises the question of what other such technologies might be lurking in our past, in our present, or, more important, in our future.

For the *Hindenburg* was by no means an isolated case of a mania-driven pathological technology. Certain similarly outsized projects—some of more recent vintage, others still being pursued today, and still others planned for the indefinite time to come— were and are motivated by the same sorts of mystic spells, emotive forces, and mind-numbing infatuations as was the zeppelin. For example, between 1957 and 1974, which is to say, for a period of seventeen years, the US government undertook, advocated, and financed an enterprise known as Project Plowshare. Plowshare's purpose, under its primary cheerleader and intellectual champion, Edward Teller, was to perform a series of planetary engineering feats on a grand scale, excavating canals, harbors, rivers, and road cuts, among other things, by detonating a succession of nuclear bombs.

In 1957 Teller and his crew, based at the University of California's Livermore Laboratory, started laying plans for the first of these exploits: a nuclear earthmoving demonstration to take place at Cape Thompson, Alaska. There a series of six atom bombs would give birth to a new harbor practically instantaneously—"in a matter of milliseconds," Teller said.

In theory, those half-dozen nuclear explosions would transform a small stream (Ogotoruk Creek) into a major international shipping destination and lead to fabulous economic development in the area. As for the radiation and fallout created by the blasts, Teller minimized their possible dangers. "We expect that all except 10 or 20 percent of the radioactive by-product will be trapped at the deep zero points," he said, "and we hope that it will remain practically immobilized in the fused rock."

But that initial proof-of-concept event would be merely the opening salvo in a series of several much more ambitious nuclear geoengineering projects to be carried out elsewhere. Plowshare scientists devised plans for H-bombing a new and improved sea-level canal across the Panama isthmus, the "Pan-Atomic Canal"; a new highway route through the Mojave Desert between Needles, Arizona, and Barstow, California; a new harbor in northwest Australia; and other instantaneous creations. Some of these ventures would require an enormous number of bombs: the Pan-Atomic Canal was estimated to require anywhere from 260 to more than 700 nuclear devices, producing a total energy release in the range of 20,000 times as much as the bomb that destroyed Hiroshima.

At its peak, Project Plowshare employed hundreds of scientists and was frittering away hundreds of millions of dollars on plans for nuclear geoengineering operations at various points around the globe. But eventually the American government came to understand that because of the massive amounts of radiation that its planned bomb blasts would release into the atmosphere, the entire program was mildly insane. It was therefore canceled, but not before several trial blasts were set off at sites in New Mexico near Carlsbad and in the Carson National Forest; others in Colorado near Rulison and in Rio Blanco County; and still another at the Department of Energy's Nevada Test Site.

It is easy for us today to recognize Project Plowshare for the lunatic folly that it was. But it didn't seem so at the time—at least not to some of the nation's best and brightest atomic scientists. What similarly delusional, pathological, arguably foolhardy technological marvels might today's scientists be hatching in their labs, conferences, and technical papers? Answering that question will require that we have a set of criteria that will define a pathological technology and allow us to separate off nonpathological, benign, and valuable uses of applied science from those that are dangerous, misguided, or, at best, gigantic wastes of money. One of the most amazing aspects of pathological technologies is their duration: they are actively pursued for years or decades, even after their harmful or destructive effects become readily apparent to all concerned. It would be useful, therefore, if we could identify such threats in the bud and perhaps avoid developing them. Knowledge of the pathology might inoculate us against its effects, or prevent us from developing the technology in the first place.

First, *a pathological technology typically (but not always) embraces something huge, either in its effects (such as a nuclear explosion) or in its absolute size (such as the zeppelin).* In either case, there is an extreme disproportionality between the mega-scale size of the technological object and the relatively puny size of the agents who are, at least nominally, in control of it—which is to say, human beings. The largest zeppelins were such gargantuan beasts that it took upwards of 200 ground handlers to pull one down out of the sky or to hold it against wind gusts, and even then, that horde of men wasn't always successful. A pathological technology is one that is barely under human control. Many zeppelins, in fact, were not.

Still, heroic size is not the core attribute of a pathological technology. More than anything else, what makes a given technology pathological is that *it exists within, is a product of, and induces a virtually paralyzing state of emotional fixation, a condition bordering on hypnotic enthrallment on the part of its proponents.* Not only is the technology motivated by such emotion, but it elicits similarly emotional reactions from its beholders.

This second feature of a pathological technology is especially striking in the case of the zeppelin. Zeppelins cast a spell upon virtually all who saw them. These machines were stately, silent, and awe-inspiring; they were not so much aircraft as small worlds.

Seeing one was like experiencing a sort of ecstatic dream vision, a solid apparition that loomed over you in the heavens, floating quietly in an ocean of air. For all of their defects, illogic, absurdities, and hidden dangers, their sheer size gave them an aura of invincibility approaching omnipotence. So, in the end, what was a little risk in the face of this cosmic object that rose above you in the sky like the moon?

The zeppelin drove Germans into a frenzy. And paradoxically, it was not its successes that inspired the greatest heights of delirium, but its failures. The most lavish public outpouring of support for Count Ferdinand von Zeppelin came in the wake of the crash, burning, and destruction of his airship LZ 4 in 1908, near the small German town of Echterdingen. Had it not been for the millions of German marks that members of the general public showered upon the Count at that point, it is likely that his airship would have died a sudden and deserved death, and the *Hindenburg* disaster never would have occurred.

That intensity of emotion, and the way it substantially blots out skepticism, criticism, and reason, explains a pathological technology's persistence across time: it endures because rational thought is virtually powerless against it. The potent force of emotion also explains the third defining characteristic of a pathological technology: *its proponents regularly and systematically underplay its downsides, risks, unintended negative consequences, and even blatantly obvious dangers.* Zeppelin Transport Company officers attempted to reassure and capture potential passengers by pointing out the strengths and virtues of their craft, their sound construction, and their safety features. They touted the peacefulness of the takeoffs, the comfortable, smooth ride even in turbulence, the splendid meals served aloft. The minor detail of traveling in what amounted to a flying bomb was passed over in silence.

As for Project Plowshare, the refusal of the program's scientists, administrators, and public relations specialists to acknowledge the risks of radiation damage was a truly impressive performance. Radiation, they maintained, would remain safely in the ground or, if it escaped, would blow away harmlessly elsewhere; in any event, it would be so marginal in its intensity as to constitute no hazard to human, animal, plant, or marine life. Fallout apparently would wind up somewhere on the Moon's Mare Tranquillitatis. Denial, magical thinking, and underestimation of any and all possible dangers are inescapable parts of the rhetoric and practice of a pathological technology.

Fourth, and finally, there is the issue of *the cost of the technology in relation to the potential benefits that it is alleged to confer.* The cost of a technology can be measured not only in dollars but also by its use of resources, by the magnitude of its damage to the environment, and by its expenditure of time and effort, as well as by other metrics such as its effect on human health or in human lives lost. In general, *a technology is pathological when its benefits are dwarfed by the scale of its relevant costs.* Because of its callous disregard of human life and health, as well as the potential extent of its damage to the environment, Project Plowshare was the hands-down winner on this criterion.

But a technology does not have to kill or physically harm people in order to be arguably pathological, even if perhaps only weakly so: the Superconducting Supercollider (SSC), a proposed fifty-four-mile-long particle accelerator to be built during the 1980s and 1990s in Texas, would have cost billions of dollars, but with possible returns that would have been restricted largely to a small and elite group of theoretical physicists. Later, in 2001, after the cancellation of the SSC, a separate group of high-energy physicists entertained the prospect of building an even larger accelerator, one that would be fully 233 kilometers, or *145 miles,* in circumference. That would enclose an area larger than Rhode Island by more than 400 square miles—almost twice as big as Luxembourg. Here there is a disproportionality between the technology's financial and environmental costs and its use of valuable resources, on the one hand, and, on the other, its likely benefits to humanity.

These four criteria—the oversize nature of the technological objects, their origin in an emotional fever, a systematic minimization of risks and possible harms, and an extreme mismatch between costs and benefits—are not mathematically precise standards, nor are they mechanical or automatic in their application. Technologies are complex and multifaceted phenomena—they are not monolithic or one-dimensional. In his book *What Technology Wants,* Kevin Kelly notes, "If we examine technologies honestly, each one has its faults as well as its virtues. There are no technologies without vices and none that are neutral." Equally, because of their variegated nature, few if any technologies are purely pathological or nonpathological. Both zeppelins and Project Plowshare's nuclear bombs conferred *some* benefits, after all, but at risks and costs that were unacceptable.

Thus, the concepts *pathological* and *nonpathological* are not neat, simple, binary, either-or categories. Rather, they exist as extremes along a continuum, with some cases falling at one endpoint or the other and other cases in between. Technologies may be strongly or weakly pathological, and there can be borderline cases, as well as cases that are undecidable in principle by any method. The problem of classifying technologies is not like deciding whether a given whole number is odd or even, a question for which there are several different objective decision procedures that give the same and mathematically correct and certain answer.

The four criteria of a pathological technology are more like diag-
nostic indicators than absolutely definitive guarantees. They are
like signs and symptoms, such as the warning signs of a heart attack: you may experience them for accidental reasons while remaining relatively healthy. Still, you should not ignore them.

Like heart attacks, pathological technologies are real and genuine phenomena—they are not imaginary. And the subject is an
important one, for human lives, states of health, environmental effects, valuable resources, and large sums of money are at stake. At
least one grand-scale enterprise even now being proposed for the
future betrays all the hallmarks of a pathological technology: an
interstellar spaceflight to be made by a Space Ark, a vehicle containing thousands of people that would travel at near-relativistic
velocities, propelled by an exotic, fantasy-level propulsion system
across light-years of space to an essentially unknown destination.
Such a craft is disproportionately gigantic, its mission bathed in
the halo glow and all the emotional allure of "going to the stars,"
with any and all costs and risks dismissed under the all-purpose
rationalization that the call to explore is an ancient and inescapable part of our DNA as a species and must be heeded no matter
what. This would be a voyage, furthermore, that provided scant
benefits to anyone except those aboard—in the unlikely event
that they, or their descendants, survived the trip. Indeed, because
of the extreme nature, variety, and abundance of the hazards it
faces, the Space Ark (or any other starship) could well be the
Hindenburg of the future.

But we want no more Project Plowshares, no more particle accelerators the size of small nations. And once we see it in the fullness of its schizophrenic extravagance and danger, we certainly
want no more *Hindenburg*s.

Prologue

"UP SHIP!"

RHEIN-MAIN WORLD AIRPORT, FRANKFURT, GERMANY
MONDAY, MAY 3, 1937
LATE AFTERNOON

Airship LZ 129, the *Hindenburg*, at that time the world's biggest aircraft, the star vehicle of the German Zeppelin Transport Company, was floating motionless in its hangar. This was an immense ship, one whose dimensions expressed numerically—804 feet long, 150 high from the top of its upper fin to the bottom of the lower one—do not begin to convey its true, almost otherworldly, size. It was so big that ordinary notions of scale completely failed to apply. If stood on its nose, the craft would have overtopped the Washington Monument by 250 feet. It was longer than three of today's Boeing 747s lined up in a row. Indeed, it approached the proportions of the monumental *Titanic*, whose width it exceeded by 43 feet.

Zeppelins were big, but they were also deeply schizophrenic flying machines. They were unbelievably, stupendously gigantic,

1

but they carried at most 72 passengers, and in many cases half that, whereas the *Titanic* carried more than 2,200. On some flights there were twice as many crew members aboard as there were passengers. Airships bombed cities during World War I, but their navigation was generally so poor that they frequently hit the wrong target—and in one case even the wrong country. Zeppelins were marvels of construction and state-of-the-art technology, but the inner gas cells that held the ship's lifting substance, hydrogen, were made out of an archaic animal membrane known as gold-beater's skin—cattle intestines, lots of them.

Airships were exquisitely vulnerable to wind and weather. Early zeppelins got ripped from their mooring masts and were dragged sideways into structures or trees or were pushed to the ground. At their best they were faster than ocean liners, but at their worst they were slow enough that in strong headwinds they sometimes hovered motionlessly over the earth or, even more embarrassingly, were actually blown backward.

It was a grand way to travel, with public areas—dining rooms, lounges, bars, promenades—being as well appointed and opulent as those of the cruise liners they rivaled. At the other extreme, however, zeppelin bedrooms resembled Pullman-car sleeping cabins or, worse, even prison cells, with cramped upper and lower berths, no toilets, and, in the case of the *Hindenburg,* only a single feeble shower ("more like from a seltzer bottle") to serve all passengers collectively, as if in a summer camp or a hippie commune. In flight, zeppelin engines and steering systems often failed, fan blades broke off, crankshafts and bearings melted, propellers departed from their mountings and spun to the ground. Landing an airship was so chancy a proposition that a commander liked to have a vast armada of ground crew members present who could hold on to the craft by means of guy lines dropped from ports in the hull and then help maneuver it toward the mooring mast. As a result of their extreme size, internal weaknesses, and other faults, the ships crashed and burned with monotonous regularity. World War I zeppelin crew members referred to their machines, quite accurately, as *Fliegende Krematorien* ("Flying crematoria").

The biggest incongruity of all, however, was the *Hindenburg* itself. The ship's immense interior was filled with more than seven million cubic feet of hydrogen, a highly flammable gas. This made it one of the most explosive assemblies ever created. Despite that circumstance, the ship attracted its share of well-heeled and apparently fearless passengers.

The *Hindenburg's* trip that night—a flight across the Atlantic to the US Naval Air Station at Lakehurst, New Jersey—would be the sixty-third of its brief one-year career. By every measure, the craft was in excellent condition, its airworthiness certificate having been renewed barely two months earlier, on March 11. Weather conditions, however, were not quite ideal for the departure, which was scheduled for 7:00 p.m., in the cool and calm of late evening. It had rained lightly in Frankfurt that morning, and it continued to drizzle on and off through the day.

Soon, the great zeppelin would be pulled from its hangar. It did not simply motor out of the building by itself, under its own power, nor was it towed out by a tractor in the manner of latter-day jetliners. It required a substantial human ground force, as many as 300 men, to haul the beast from its lair.

The ship emerged slowly, at the rate of approximately 100 feet per minute, meaning that it took about eight minutes for it to fully exit the hangar. First came the bulbous, pointed nose, then the shiny silver forward section, followed by the letters HINDENBURG, high up on the hull and painted in a bright red Gothic script. Then, attached to the keel of the ship, the control gondola, a relatively short structure, only thirty feet long, with a large pneumatic tire at its base to cushion the shock of landing. Then came the passenger section, visible as large Plexiglas windows that were set at an angle to the hull, giving those lining the port and starboard promenade decks a commanding view of whatever lay below.

Farther back, on the same level as the ship's name, there had once had been the five interlocking rings of the Olympics. They had commemorated the 1936 Berlin summer games, at which the craft had made a flyover appearance. But the Olympic symbol had been removed during the winter. That painted-over section was now followed in turn by the forward engine car with its gigantic twenty-foot-diameter propeller, itself almost two stories high. The ship's registration code came next, "D-LZ129," rendered in large, square black lettering. Then the rear engine car, and after that a long, blank stretch of the ribbed hull until, finally, the tail end, consisting of horizontal and vertical stabilizers, came into view. On each side of the two vertical fins there was an enormous black swastika. Each stood out clearly against the surrounding white disk that in turn was enclosed by a scarlet rectangle. The Nazi Party emblem was by far the most vivid and indelible single feature of the ship.

Still, even that bold symbol was not quite the end of the parade: the rudders—huge structures themselves, each of them more than thirty feet long—as well as the elevators came next, followed at last by the gently rounded, pointed tail cone. But that, at last, was it: the *Hindenburg* in full.

Three days later, it would be gone.

At about the same time that the *Hindenburg* was being pulled from its hangar, the passengers for the flight—only thirty-six of them, although the craft had accommodations for double that number—were being picked up at the front of their luxury hotel, the Frankfurter Hof, by three airport buses. One of the travelers, who apparently had been celebrating his impending departure with copious amounts of alcohol, sang all the way to the airport, which was a half-hour ride away.

The buses contained a prosperous crowd. A trip on the *Hindenburg* was so expensive that only the world's elite could afford it. Passengers traveled in almost decadent opulence, at least in the public areas. The craft boasted practically every amenity of first-class airline or ocean liner travel—except for the sleeping compartments, which were narrow, with exceptionally thin walls between them. In the dining room the passengers would be treated to chef-cooked gourmet meals, fine wines, silver serving plates, printed menus, and fresh flowers on every table. The *Hindenburg* had a club lounge, plus a small bar, and a separate reading and writing room. And however incongruously, given the proximity of the sixteen hydrogen gas bags overhead, the craft also had a hermetically sealed and pressurized smoking room with an airlock door that was closely monitored by a crew member. That was on the lower level, which was also called B deck. Above, on A deck, a pair of 200-foot-long promenades ran front to back, one on each side of the ship. The large viewing windows could be propped opened even in cruising flight.

The passenger list for the trip ranged from children to company presidents and included journalists, businessmen, salesmen, executives, a former Luftwaffe pilot, a spinster American heiress living in Rome, a family of four from Mexico, and an acrobat, among other people. The most common motive for them to be traveling aboard the *Hindenburg* was a desire to get to the States fast. "*Traversez l'Atlantique en 2 Jours*" read a French version of the German Zeppelin Transport Company's advertising poster, a bit optimistically, as opposed to six days or a week by steamship.

One of the passengers was Marie Kleeman, wife of a German motorcycle manufacturer. She was traveling alone, her final destination Amherst, Massachusetts, where she was to visit her daughter, who was recovering from surgery. Mrs. Kleeman was one of eight women aboard the *Hindenburg* that night, including Emilie Imhof, who was not a passenger but part of the crew, a stewardess. Of the eight, five would survive and three would either be killed in the wreck or die shortly afterward, in the hospital. At the time of the fire, Mrs. Kleeman would be seated at a window of the port-side promenade.

There was the Doehner family: Hermann and Matilde, together with their children Werner (age eight), Walter (ten), and Irene (fourteen). Hermann was the general manager of a wholesale drug company with head offices in Mexico City, to which the family was returning after a trip to Germany. One of the family would die in the wreck, another in Point Pleasant Hospital. Of the three who survived, one is still alive today. He is in fact the last living survivor of the *Hindenburg* disaster.

There was Nelson Morris, an Armour & Company meatpacking executive, who was probably the wealthiest man aboard the ship. He held a pilot's license and loved airship travel. After his first flight on the *Graf Zeppelin,* he had written to his brother Edward: "This is the most interesting experience I have ever had in my life. . . . I cannot find the words to properly express the sensation." He was traveling with Burtis "Bert" Dolan, who had previously worked for Armour too but was now vice president of the Lelong Importing Company, which dealt in perfumes. Dolan had planned to return home from Europe by ocean liner, but Morris persuaded him to fly on the *Hindenburg* instead, this despite the fact that Dolan had solemnly promised his wife that he would never fly. By taking the *Hindenburg,* Morris told Dolan, he could make it home by Mother's Day on the ninth, to the surprise and delight of his wife.

The heiress was Margaret Mather, who had been born in Morristown, New Jersey, in 1878, but had lived in Rome since 1906. She was now fifty-eight and was making her first trip back to the States in eight years, to see her brother Frank, emeritus professor of art and archaeology at Princeton. She had one of the new cabins on the lower deck of the *Hindenburg,* one of the few with windows; the rest of the cabins, upstairs on A deck, were inside rooms. As the only unattached adult female passenger, she would be seated at the Captain's table during meals, next to Max Pruss, who was in command of the flight. At the time of the crash, Miss Mather would be in the port-side promenade, just steps away from Marie Kleeman.

Peter Belin was a twenty-four-year-old resident of Washington, DC, now returning home from a year at the Sorbonne. Like Nelson Morris, Belin had an avid interest in flying and was a licensed pilot.

The acrobat was Joseph Späh, a vaudeville performer, stunt man, comedian, and wit, who was on his way home to his family in Douglaston, New York, after a long European tour. He was accompanied by his performing dog Ulla, an Alsatian Shepherd, who was kenneled in one of the cargo spaces at the rear of the ship. During the voyage, Späh would pay frequent visits to Ulla, to feed and walk her along the keel walkway. He also had with him a movie camera and at one point would shoot a passing iceberg off the coast of Newfoundland. He would also capture a ghostlike image of the *Hindenburg*'s shadow moving slowly across the face of the ice. Incredibly, a portion of the film survived the fire and a short clip from it—eerie, silent, and in black and white—can be seen to this day on YouTube.

Helmut Lau was one of the *Hindenburg*'s crew members, a helmsmen, of which there were three on board. At the time of the fire, he would be standing on a ladder leading down into the auxiliary control room at the very bottom of the lower vertical fin. He would hear a muffled detonation, look up to find its source, and see a bright light at the front of hydrogen cell number 4. What produced that light would be the subject of debate for a long while after the event.

And then there was Werner Franz, fourteen years old at the time of the flight. As the *Hindenburg*'s cabin boy, it was his job to serve meals to officers and crew members and to clean up afterward, which he did in the officers' mess and kitchen, which were located on the lower deck. He was at this point practically an old hand aboard zeppelins, having already made three flights to South America on the *Hindenburg* the previous year. This would be his first visit to the United States, and he was looking forward to seeing New York City.

Whatever their background, and irrespective of whether they were passengers or crew members, each would suffer his or her own separate fate: some would be incinerated by the fire within minutes; others would be covered with burned flesh and die a horrible, painful, lingering death; still others would walk (or run) away from the blaze and emerge with barely a scratch or even so much as a singed eyebrow.

In all, the thirty-six passengers and sixty-one crew members on that night's flight made for a grand total of ninety-seven souls on board, including no less than six captains of the line.

The commander of the ship, Captain Max Pruss, had made his first flight aboard an airship (the Naval Zeppelin L3) in 1914, and he had flown in the German navy's zeppelins during World War I. He had joined the Zeppelin Transport Company after the war and worked his way up the ranks until finally becoming commander of the *Graf Zeppelin* in 1934. By the time of the *Hindenburg's* final flight, Pruss had more than twenty years of experience in airships behind him, and he was as qualified as any of his peers to be the commanding officer that night. Still, in the aftermath of the disaster, he would be regarded as having made some extremely questionable judgments during the craft's final moments.

The five other captains were Albert Sammt (first officer), Heinrich Bauer and Walter Zeigler (watch officers), Ernst Lehmann (adviser), and Anton Wittemann (observer). If anything, there would be a wretched excess of Zeppelin Company top brass aboard the flight. There was a reason for this, but it would be explained only later.

When they arrived at the Rhein-Main World Airport, crew members asked the passengers to hand over any matches or lighters, which would be placed in numbered bags that would be returned to them at the end of the trip. (The smoking room was equipped with a single electric lighter that was to be used by all the smokers aboard.) Then they boarded the ship, by means of two retractable aluminum staircases that ascended to the lower level, from which two larger, internal stairways rose to a common landing area on A deck. There, on a pedestal inside a small niche was a bronze bust of the late Field Marshal Paul von Hindenburg, after whom the ship was named. Stewards conducted passengers to their "staterooms," their little cubbyholes furnished with collapsible washstands, fold-up writing tables, and upper and lower bunk beds.

It didn't take most of the travelers very long to escape from their mostly windowless cabins and gather together on the port and starboard promenades to watch the takeoff ceremonies. Not that these amounted to much. Arrayed out on the tarmac on the port side of the airship was a brass band whose members were dressed in blue-and-yellow uniforms. They played some bright military tunes, but then ended with *"Ein feste Burg"* ("A Mighty Fortress"), the hymn composed by Martin Luther.

The launch procedure was simplicity itself. At a word from the commander, a bunch of ground crew members who were holding on to the grab rails of the control car would give the ship an upward push, after which the *Hindenburg,* an 804-foot-long aerial monster, would rise up into the air as if it were no heavier than a puff of smoke.

So, at a little after 7:00 p.m., on Monday, May 3, 1937, Captain Max Pruss, master and commander, gave the time-honored, almost thrilling order: "Up ship!" And the *Hindenburg* floated up into the oncoming darkness.

PART I

BEGINNINGS

Chapter 1

THE MAN IN THE SKY

On May 6, 1863, at the height of the American Civil War, a twenty-five-year-old cavalry officer from the independent German kingdom of Württemberg arrived in New York City aboard the Cunard steamer *Australasian*. He was thin and dapper and had a bushy beard, long sideburns, and a mustache.

This was Count Ferdinand von Zeppelin, who had come to the United States to observe the course of the fighting. As he'd written in a letter to his sister Eugenia, "I dare to hope that war, which I have chosen as the main study of my life, might be revealed to me in its bloody truth."

He traveled to Philadelphia, where he stopped for a while, and then went on to Washington and took a room at the Willard Hotel, which even then was considered a deluxe fixture of the US capital city. Franklin Pierce had stayed there before his inauguration as president ten years earlier, in 1853. The Prussian ambassador soon arranged an audience with President Lincoln, and for his visit to the White House Zeppelin fitted himself out in a morning coat and top hat. He was escorted into the president's office, where "there rose from behind the desk a very tall spare figure with a large head and long untidy hair and beard, exceptionally prominent cheek-bones, but wise and kindly eyes."

Zeppelin's purpose in seeing the president was to gain permission to travel with the Union armies, for he wanted to experience firsthand all the heady joys of combat. In due course Lincoln authorized a pass that allowed Zeppelin to travel with and move freely among the armies of the North. Thereupon he acquired a horse that came equipped with an American wooden saddle, which was different from the English saddles he was accustomed to back home. He also hired a black servant named Louis.

Soon he was attached to the staff of General Joseph Hooker's Army of the Potomac, which was then encamped near Falmouth, Virginia. Here, as Zeppelin said, he "was introduced to the romance of war the very first day. It was an awe-inspiring experience for me."

Zeppelin, as we would say today, was "embedded" with Hooker's army for about a month. Despite his official status as a noncombatant, he was nevertheless very much a man of action, and so he often made night rides, carried secret dispatches, and took part in reconnaissance missions. In June 1863, he was present at Ashby's Gap as an observer of a battle in which Hooker's forces lost to those of the famed Confederate general J.E.B. Stuart. Lincoln relieved Hooker of his command shortly afterward, at which point Zeppelin also left the Army of the Potomac.

He did not think much of the Union army at any rate. As he wrote to his father in somewhat exaggerated terms, there was "no systematic cooperation, no local patrol work, no enemy intelligence, no general staff, no maps, no corps combining all the different arms, no tactics adapted to local topography."

So much for America's military greatness. Zeppelin decided that he might as well see the country while he was there, and so he set off on an expedition whose purpose was ostensibly to discover the source of the Mississippi. The trip included visits to "the magnificent Niagara Falls," Lake Erie ("one of the smaller of the group of lakes"), and the Mississippi River itself ("rather wide but very shallow"). By August, after a somewhat harrowing overland and canoe journey with two Russian companions and two Indian guides during which all of them nearly starved to death, he wound up in St. Paul, Minnesota, where he checked into the International Hotel under the name of "Count Reppelin" or "Zeppelerlin," according to competing local newspaper accounts of the day.

Across the street from his hotel, a barnstorming balloonist by the name of John H. Steiner, a German who lived in Philadelphia, was giving sightseeing rides in a 41,000-cubic-foot balloon fired by coal gas, for the price of $5 per passenger. The contraption had previously been used as an observation platform by the Union army, which had its own Balloon Corps. On August 19, 1863, Count Zeppelin made his first-ever balloon ascent, riding in the basket "to an altitude of six or seven hundred feet."

From his vantage point in the sky, Zeppelin could see the city of St. Paul, the Mississippi, and, to the west, a ridge of hills that lay parallel to the river. Ever the military thinker, his first thought was that this airborne basket would be an excellent command post from which to direct firepower against an enemy. As he wrote to his father: "Should one want to harass with artillery fire the troops deployed in reserve on the other slope, the battery could be informed by telegraphic signals where their projectiles hit. . . . No method is better suited to viewing quickly the terrain of an unknown, enemy-occupied region."

By rights, this balloon ascent ought to have been the turning point in the life of the inventor of the first successful rigid airship. There he was, a speck in the sky, in a lighter-than-air vehicle, surveying an imaginary battlefield and overlooking all of creation. That just *had* to be the beginning of his lifelong crusade to invent and perfect the dirigible. And indeed, he himself later said as much, although not until a full fifty-two years after this supposedly pivotal event. In an interview with an American journalist, Karl H. von Wiegand, in 1915, two years before his death, Zeppelin said: "While I was above St. Paul I had my first idea of aerial navigation strongly impressed upon me and it was there that the first idea of my Zeppelins came to me."

Historians of the airship, however, have tended to dismiss this as a false memory. Even Hugo Eckener, Zeppelin's longtime business partner, friend, advocate, and, later, biographer, wrote in 1938 of the famous St. Paul balloon ride: "It is not . . . true that he was already at that time thinking about airships or that this ascent suggested the idea to his mind."

But if it didn't, then what did?

Zeppelin, according to Eckener, was a "man obliterated by his achievement." And indeed, while Ferdinand von Zeppelin was arguably the major driving force behind his eponymous creation, many people know little or nothing about the man himself.

He was born on July 8, 1838, in Germany, in a building that was once a Dominican monastery and is now the Steigenberger Inselhotel Konstanz, located on a small private island in Lake Constance (also called the Bodensee). The island was, and still is, connected to old-town Konstanz by a wooden footbridge. Zeppelin's was a family of landed and moneyed nobility, and his world was not far removed, in spirit, from the feudal system of the late Middle Ages. There were neither knights nor vassals in that world, but even in the midnineteenth century a nobleman's loyalty was to the king, and one's primary concern was a romantic devotion to the arts of war. That attitude suited Zeppelin to a tee, for he was a soldier not only by profession but by temperament: throughout his life he considered himself a dyed-in-the-wool military man. He was emotionally in tune with George Washington's observation that "I have heard the bullets whistle; and believe me, there is something charming in the sound."

Zeppelin grew up on the family estate at Girsberg, Switzerland, where for a while he was schooled by a private tutor by the name of Robert Moser, a prelate. In 1853, when Ferdinand was fifteen, his father, a member of the Württemberg aristocracy, sent him first to the Polytechnic School in Stuttgart, and then two years later to a military college in Ludwigsburg. Upon graduating, Zeppelin began his formal career as an officer in the army of the King of Württemberg. Not long afterward, however, he asked permission to attend the university at Tübingen, where for a short time he studied political economy and history plus a smattering of chemistry and mechanics—his only formal training in the sciences.

No bookworm, Zeppelin viewed himself as very much a practical, no-nonsense, hands-on man, and so in 1861 he took off on a grand tour of Europe to immerse himself in the military traditions of other countries. He went to Vienna, where he met the Emperor Franz Joseph, and then to Trieste, Venice, Verona, Genoa, Marseilles, and, of course, Paris. He then traveled to the northern countries, including Belgium, England, and Denmark. Having missed hardly a single country in Europe, he at that point left for the States.

After his return, Zeppelin was appointed adjutant to King Charles of Württemberg, in whose service he took part in a number of minor battles. It was during some of them that Zeppelin acquired a reputation—at least within Württemberg military circles—for being headstrong and determined and for not always using the best judgment. At the Battle of Aschaffenburg, for example, Zeppelin was charged with delivering a message across the River Main, which was then quite swollen and running a strong current. Clad in riding boots and full military regalia, complete with sword, he swam across the river, nearly drowning in the process, and then also swam back, arriving on the opposite shore in even worse condition. He was, one might say, slightly unorthodox in his habits and methods.

Later, in 1870, during the Franco-German War, Zeppelin left on a reconnaissance mission accompanied by a dozen fellow cavalrymen and rode some twenty-five miles behind enemy lines on a succession of ever-weakening horses. "My horse had lost a lot of blood from a wound and I replaced it by a police mount, but when the latter refused to jump the ditches, I had to take over a still smaller horse," and so on. As it was, with all of his comrades killed or lost, he returned from the exploit alone, a feat that earned for him the labels "daring," "dashing," and (in the words of Field Marshal von Moltke) "a hothead."

Still, at this stage in his life everything pointed to Zeppelin's finishing out his beloved military career at a ripe old age and in a blaze of glory. And that would likely have been his fate had he not made his One Big Mistake. A proud Württemberger, Zeppelin had long objected to the way in which members of the Prussian aristocracy lorded it over the officers of this independent state of the German Empire. "Württemberg officers still found the Prussians hard to accept," wrote one historian of the period. "They were arrogant, tactless, and introduced an unaccustomed flashy lifestyle into quiet Württemberg."

In early 1890, in an effort to correct the situation, Zeppelin wrote a "secret memorandum" to the German Kaiser, Emperor Wilhelm II, in which he set forth his heretical views and complaints: he told the Kaiser, in so many words, that the practice of Prussians having command over Württemberg officers "has no justification either in the imperial constitution or in the military convention"; furthermore, he claimed, such a policy turned the King of Württemberg into "a mere rubber stamp."

Zeppelin in effect was telling the emperor how to run his own empire, a rather naive piece of effrontery with which Wilhelm II was not amused. "Much surprised to read these particularist ideas!" the Kaiser wrote in the margin of Zeppelin's memorandum. (By "particularist" he meant a subordinate state's assertion of its individual rights within the greater realm.)

It was therefore quite understandable—except perhaps to Count Zeppelin—that by the end of the year the Kaiser had "promoted" him into retirement, using as a pretext the claim that the Count had done poorly in his latest regimental exercises. This action put Zeppelin in a decidedly bad light among his peers, for it amounted to a public disgrace. In addition, he was now, at age fifty-two, essentially a fish out of water—and a man without a purpose.

It was an altogether demoralizing experience. In a study entitled *Count Zeppelin: A Psychological Portrait,* the historian Henry Cord Meyer claims that Zeppelin's humiliation "caused him extreme psychological trauma. . . . Unless he found some alternative and effective way of self-expression, he was doomed to enforced retirement and social-psychological deterioration on his Swabian estate."

But if his separation from the army was an emotional tragedy, Zeppelin's recovery from the experience was a national disaster, for it would lead, in a roundabout way, to his going on to invent the pathological technology of the hydrogen airship.

In 1890, Zeppelin's *annus horribilis,* his future partner, Hugo Eckener, was twenty-two years old and attending philosophy courses at the University of Berlin. Unlike the Count, Eckener was not of the nobility, nor did he cut the dashing figure that Zeppelin did: Eckener always tended to look slightly rumpled and messy, like a half-shorn sheep. He grew up in northern Germany, in the Baltic seaport town of Flensburg, where his father Johann owned and operated a tobacco, cigar, and cigarette factory. Eckener's birthplace in Flensburg, a building at Norderstrasse 8, is today the Eckener Haus Restaurant, and a model zeppelin hangs over its doorway.

Eckener and Zeppelin did have one thing in common, and it would turn out to be a major factor in their ultimate success in building, flying, and navigating airships. Both had grown up near large bodies of water, and since both loved the sea and sailing, they were exquisitely sensitive to the elements, to the errant ways of wind and weather.

Whereas Zeppelin pursued his military career single-mindedly, as if it were a divine calling, Eckener's interests were more wide-ranging and mercurial, and he always seemed to be departing simultaneously in all directions. Before pursuing philosophy in Berlin, he had studied logic, aesthetics, history, medieval literature, and the history of art at the university in Munich. Then, while in Berlin, he decided that there was no future for him in the philosophical profession and he needed to enroll in something more practical. So he transferred to the University of Leipzig, where he took up experimental psychology, studying with Wilhelm Wundt, who had more or less invented the discipline.

Eckener got his doctoral degree in psychology, *magna cum laude,* with a thesis on "Variabilities in Human Perception." On the strength of this achievement, he was offered a position at the University of Toronto, where he was asked to create and direct an Institute of Psychology. Eckener declined this honor, however, and instead joined the German army in order to complete his required year of military service. That done, he took a somewhat lowly job as editor and writer for the *Flensburger Nachrichten* (the *Flensburg News*), his hometown newspaper. Meanwhile, he began thinking about doing something more ambitious and was planning to write a book about economics.

In 1896, at the age of twenty-eight, it was almost as a respite from all of this breathless and knockabout intellectual activity that Eckener proposed (by mail) to Johanna Maass, the sister of a friend of his. Joanna accepted, and they got married the next year. It must have been quite a relief for Hugo Eckener.

At least momentarily. They settled at first in Munich, but soon moved to the rustic town of Friedrichshafen, which was practically as far south as you could go in Germany and still be in the country. The chief attraction of the place for Eckener was that it was on the northern shore of Lake Constance, on whose waters he could renew his lifelong passion for sailing. As it happened, Friedrichshafen was also the place where, just before the turn of the twentieth century, Count Zeppelin would start building and flying—and crashing—his airships.

Neither of these men, who would jointly bring the rigid airship into existence, had any formal training in physics, engineering, or aeronautics. They started off as rank amateurs, became experts, and invented a flying machine that was fatally flawed.

Even before his traumatic fall from grace in 1890, Zeppelin had been ruminating about air travel. Fully twenty years earlier, in 1870, seven years after his fabled balloon ascent in St. Louis, Zeppelin had been present at the Siege of Paris, when the German army had surrounded and blockaded the city. To maintain communications with the outside world, the Parisians took to using carrier pigeons and hot-air balloons, which during the four-month siege had transported more than a million pieces of mail (some of which had been photographically reduced to something like microdots in order to save weight) at the postal rate of 20¢ per letter. A few of the balloons had also carried people, and Zeppelin had been highly impressed by the escape from the city of Léon Gambetta, a future prime minister of France, aboard a balloon named the *Armand-Barbès*.

Incredibly enough, Gambetta actually arrived in Tours, his intended destination, but balloons were so much at the mercy of the winds (unlike carrier pigeons, which followed their inbred homing instincts) that you could not depend upon them to take you exactly where you wanted to go. One of the more than sixty balloons that flew out of the city during the Siege of Paris, the *Ville d'Orléans,* left on November 21, 1870, and did not touch down until fifteen hours later, when it landed near Christiana (today's Oslo) in Norway, setting a new distance record in the process. If only these things could be *steered,* thought Zeppelin, they could be of practical use in wartime.

In 1874, Zeppelin's budding thoughts on the subject were further stimulated after reading a booklet entitled *World Post and Air Travel* by one Heinrich von Stephan. Stephan was at the time Germany's postmaster-general (he was also the inventor of the postcard), and he was a zealot about delivering the mail in the speediest possible fashion. In this pamphlet, which was based on a speech that he had given before the Scientific Society of Berlin, Stephan declared that "Providence has surrounded the entire earth with navigable air. This vast ocean of air still lies empty and wasted today, and is not yet used for human transportation." Nor was it used for mail.

But Zeppelin, with the Siege of Paris in mind, knew otherwise. Now, laid up and recuperating from a riding accident, he sketched out in his diary entry for March 25, 1874, a plan for a powered and steerable balloon. Under the heading "Thoughts about an airship," he described the device in three broad conceptual strokes.

First of all, the thing would be *big:* "The craft would have to compare in dimensions with those of a large ship," he wrote. Second, it would be suspended in the air by a gas, and "the gas compartments would be divided into cells which can be filled and emptied individually." Third, its ascents and descents would be controlled aerodynamically by means of movable planes that would work by action and reaction against the oncoming airstream. "The ascent will then take place through forward motion of the machine, which will force the craft so to speak against the upward-inclined planes. . . . To descend the surfaces will be angled even less upward, or speed will be reduced."

What he described, then, was a compartmented gas balloon that was aerodynamically driven and as big as a ship. Zeppelin's diaries have not been made public, and accounts differ as to whether at this stage he said anything about the shape of the craft, whether it would have a rigid or nonrigid framework, or exactly how the ship would be powered; efficient internal combustion engines were still in the future. He did, however, speak of the craft as carrying passengers, whose "parachutes, if they can be used at all, could be attached as part of the ceiling of the passenger compartment."

It is a surprise, given his earlier history, that at this point Zeppelin conceived of the airship not as a military vehicle but rather as a craft for use as "a luxury sport" or for use "where land or water communications between two points are especially difficult and call for permanent easy, safe, and rapid means of transit." But he was far from committed to any one design or overall plan; it seems that he was really just trying out ideas. In a later diary entry, dated November 29, 1877, Zeppelin rejected his initial movable-plane

scheme and asked himself instead: "Should not ascent and descent be operated by two screws [propellers] on a vertical axis?
The wings (for dynamic movement) could then be dispensed
with." Later still, in 1878, he speculated about what the airship's
gas envelopes could be made of: "Chinese silk, very light and, if
varnished, almost entirely gasproof." (As it turned out, zeppelin
gas cells would typically be made of goldbeater's skin.)

It is a mistake, therefore, to conceive of Count Zeppelin, as he
has often been portrayed, as a far-seeing inventive genius whose
idea for a rigid, cigar-shaped airship sprang into his mind like a
bolt from the blue, full-blown and substantially complete. To the
contrary, as airship historian Guillaume de Syon has said, "Count
Zeppelin was but another inventor with no technical training, one
of hundreds claiming to have solved the problem of dirigibility."
He was an amateur who added bits and pieces to his airship idea
only when he had the time and inclination to do so.

Zeppelin was still in the army at that time, and he was merely
dabbling in the black art of airship design more or less as a hobby.
But all that changed in 1884, when he suddenly got news of a
major aeronautical advance. The French, it seemed, had already
invented the dirigible.

The craft in question, named *La France,* was the product of
two Frenchmen, Charles Renard and Arthur Krebs. In 1878
Renard, a military engineer who was a graduate of the École Polytechnique, made a study of the different ways of powering what
was essentially an elongated and streamlined nonrigid balloon.
He considered steam power, compressed air, and electricity and
wanted to experiment with all three. He applied to the army engineering corps for funding but was refused on the grounds that the
idea was impractical. Renard therefore turned to one man who

had already made a highly successful balloon flight, none other than Léon Gambetta, who had escaped the Siege of Paris some eight years earlier. Gambetta, who was now the president of the budget committee of the French Chamber of Deputies, granted Renard 400,000 francs for further research.

Renard and fellow engineer Arthur Krebs were inspired by what amounted to a world's fair for electricity and the electrical industry hosted by Paris in 1881 to create a battery-powered electric motor for use on an airship. They jointly designed a 165-foot-long balloon that would be 27 feet in diameter. (By comparison, a Boeing 707 is 153 feet long, with a fuselage of about 14 feet in cross-section.) The envelope would be filled with some 66,000 cubic feet of hydrogen, and the craft would be pulled through the air by a single immense, two-bladed wooden propeller located at the front of the craft. The propeller would be 23 feet in diameter (bigger even than those of the *Hindenburg*) and turn at the rate of 46 revolutions per minute—which is to say, at less than one revolution per second.

This gigantic airscrew would be driven by an 8.5-horsepower electric motor designed by Krebs, which would be run by an array of batteries designed by Renard. The batteries alone weighed 1,232 pounds, while the motor weighed another 210, for a combined total weight of 1,442 pounds, all of which (plus propeller shaft, gears, and various other devices, along with the two aeronauts themselves) would be carried in a 108-foot-long control gondola suspended beneath the balloon by a network of cables. The craft would be steered laterally by means of a rudder and vertically by a sliding weight that could be shifted fore and aft to lower or raise the nose.

By every measure and in every dimension, this was a contraption too big, heavy, and unwieldy to succeed. Nevertheless, it did.

In 1883 Renard and Krebs started construction in "Hangar Y" at the government balloon factory at Chalais-Meudon, about five miles southwest of Paris. The next year, on August 9, 1884, at about four in the afternoon, *La France* rose into calm air from the Chalais-Meudon parade grounds and motored off in a southwesterly direction at about fifteen miles per hour. The craft traveled through the air with "faithful obedience to the slightest indication of its rudder," according to an account submitted to the French Academy of Sciences. It made a half-circle turn at midcourse and then arrived back at its starting point twenty-three minutes later, having covered a distance of about five miles in the interim.

It was no mean feat. In the italicized words of aviation historian Richard Hallion, "This constituted an accomplishment of truly momentous significance, for Renard and Krebs had demonstrated the *first completely controlled, powered flight of any sort* from the moment of takeoff to the moment of landing, in all of human history."

Nor was this an isolated, one-off flash in the pan: Renard and Krebs piloted *La France* on six more flights in 1885, some of them over figure-eight courses, and carried passengers other than themselves, including the actress Gaby Morlay, the first woman to fly in an airship. Indeed, their craft was far more successful in every way than the first few airships that Count Zeppelin would go on to build.

The importance of *La France* was not lost on Zeppelin, who now had nightmarish visions of Germany being outstripped in aeronautics by its longtime enemy. He made notes and sketches for a ship that would easily outdo *La France:* if *La France* was big, Zeppelin's airship would be *Kolossal*. More than ten times the volume of the French ship, the Count's craft would have a gas capacity of 706,200 cubic feet. That would enable it to carry twenty passengers plus cargo and mail.

Moreover, he also decided that it was now time to notify the authorities. In 1887 Zeppelin wrote to the King of Württemberg, saying that "the airship *La France* of Captains Renard and Krebs . . . has shown the undeniable possibility of controlled flight. For airship flight to be really useful for military purposes, it is necessary that the ship should be able to make progress against strong winds, and also should not have to land except at long intervals (at least 24 hours), in order to carry out long reconnaissance missions. It should have a large carrying capacity in order to carry personnel, cargo or explosive shells. All three requirements demand a very large gas volume, hence a large airship."

Three years later, in 1890, Zeppelin was out of the army, and "with nothing better to do," as he said, he launched a campaign to design, build, and fly a craft such as he had described. The process would take him a full decade and would be marked by any number of false starts, hesitations and retreats, incredible blunders, crises, abject failures, and repeated miraculous resurrections from the dead.

The first "zeppelin" to take to the air, the LZ 1 in 1900, was scrapped after three flights. The second, in 1906, was blown away by the wind, driven into the ground, and destroyed by a storm. At the time Count Zeppelin said, "I shall build no more airships." But he did.

Another one, the LZ 4 of 1908, crashed, exploded, and burned in one of the gaudier of the many *Hindenburg* "prequels." The thing was a *Katastrophe* in every respect. But the Count always had a way of coming back from any and all defeats and winding up, most of the time, in an even better position than where he had started. He seemed to be the very epitome and personification of Nietzsche's oft-quoted dictum, "What does not kill me makes me stronger." And so, in the wake of the LZ 4 disaster, Kaiser

Wilhelm II, Emperor of Germany, the very same Supreme War Lord who in 1890 had dismissed Zeppelin from the army, now personally traveled to Friedrichshafen, conferred upon the Count the highest Prussian honor—the Order of the Black Eagle—and pronounced him, without irony, "the greatest German of the twentieth century."

Chapter 2

THE PHILOSOPHER'S
STONE OF FLIGHT

Whether or not Count Zeppelin was "the greatest German of the twentieth century," it is clear that he was not the world's first airship designer, nor was he even the inventor of the first rigid airship, as is commonly believed. For that matter, Reynard and Krebs were not the first airship designers either—far from it. The first designer of a lighter-than-air craft would have been forced to solve the major problem associated with such a vehicle, namely, finding a substance that would lift human beings off the ground and into the atmosphere, and it was neither Count Zeppelin nor even Reynard and Krebs who solved that problem.

The very existence of such a substance was implausible: it would be something that had negative weight, which seemed like a physical impossibility or a contradiction in terms. Air was already so light and insubstantial that it was not even visible. Further, the sky itself was made of air, so how could there be anything lighter than that? The search for the stuff that possessed this unlikely

floating quality was analogous to the alchemist's quest for the philosopher's stone, the mythic element that would turn base metals,
such as lead, into gold. In the case of lighter-than-air vehicles, the
quest was for a substance that would magically lift objects skyward, a "philosopher's stone of flight."

In the year 1670, an Italian mathematics professor hit upon
what is without a doubt the all-time greatest, most imaginative
and ingenious solution to the problem ever given: the substance
he proposed for lifting mankind off the earth and into the sky was
nothing less than . . . nothing itself—that is, a vacuum. This was
the theory of one Francesco Lana de Terzi, a Jesuit priest who
is regarded by some as the Father of Aeronautics, although he is
more properly to be considered as the Father of the Airship.

Unlike Count Zeppelin, Terzi was a scientist. Born in Brescia,
Northern Italy, in 1631, Terzi became a professor of physics and
mathematics at the University of Ferrara. Even though he was
a cleric, he was far more interested in science than in religion,
and he spent most of his life researching and writing a book that
was a comprehensive survey of the scientific knowledge and technology of the day. The book separated what was based upon experimental fact and logical proof from the authoritative-sounding
but essentially groundless opinions that had been handed down
uncritically through the ages. Among the technologies Terzi described in this work were a system of writing for the blind similar
to Braille, a sewing machine, and long-distance communication
by means of cannon shots.

In 1670 Terzi published a preliminary sketch of the book in
a shorter précis, or summary, entitled *Prodromo,* a subsection of
which was devoted to his "aerial ship." Here he revealed that "I
have found the manner of making a machine lighter in itself than
air, so that not only will it float on the air by its own lightness, but
it may also carry men and any other required weights."

This was a true and proper "ship" inasmuch as it consisted of a boat complete with mast, sail, oars, and rudder, but there the similarity ended, for this was not a sailing boat but a *flying* boat. The scheme by which it could be made to ascend into the air rested on a number of assumptions, the first of which encapsulated one of the secrets of lighter-than-air flight: "I will, first of all, presuppose that air has weight." As important as that was, the insight was not by itself something new: the discovery that air has weight is usually attributed to Evangelista Torricelli, who established the fact in 1643. But it was Terzi who made use of that assumption for the practical purpose of lofting a ship into the air.

Next, "I assume that any large vessel can be entirely exhausted of all, or, at any rate, of nearly all, the air contained therein." In fact, Terzi knew from accounts of the vacuum pump invented by the German physicist Otto von Guericke that it was possible to extract virtually all of the air from a globe. The resulting vacuum-filled ball would be capable of some amazing feats. Guericke's "Magdeburg hemispheres," for example, were two bronze hemispheres joined together at their rims with an airtight seal. In 1657 Guericke had withdrawn the air from the joined halves by means of his vacuum pump and then shown that a team of sixteen horses, eight attached to each hemisphere and pulling in opposite directions with all their strength, could not separate the two. But when he reintroduced air into the hemispheres, they immediately fell apart by themselves.

It followed from these and certain other assumptions that when a globe of a certain size is exhausted of the air inside it, it will rise up through the atmosphere in the same manner, and for the same reasons, that a bubble rises through a fluid such as water: because it is lighter than the surrounding medium, a principle established by Archimedes. Thus, as Terzi himself put it, "Presupposing all these things, it is certain that one can construct a vessel of glass or other material which could weigh less than the air contained therein; if, then, one exhausted all the air . . . this vessel would be lighter in density than the air itself, and, therefore, it would float on the air and ascend."

Terzi did not content himself with mere abstract theory but proceeded to furnish some actual, real-world design specifications. His aerial ship was to be suspended aloft by four large hollow balls made of "copper beaten out thin." These copper globes, he estimated, would be fourteen feet in diameter, with walls that were two to three millimeters (slightly more than one-sixteenth of an inch) thick. A mathematician and physicist, Terzi then calculated the circumference, the total surface area, and the weight of such a sphere, which he said would be 154 pounds. From his own experiments on the weight of air, modeled on those of Torricelli, he also found the volume of the sphere and the weight of the air it enclosed, which he said would be 179 pounds. Since the copper sphere itself weighed only 154 pounds, whereas the weight of the air inside it was 179 pounds, "it is evident that on exhausting the air from it, not only will it ascend, but it will also be enabled to lift a weight of 25 pounds."

This, then, was the structural core of Terzi's "aerial ship." From there it was simply a matter of scaling things up: copper spheres could be made larger and larger, and several of them could be gathered together to provide ever-increasing amounts of lift, and so "from this it can be easily seen how it is possible to construct a machine which, fashioned like unto a ship, will float on the air."

Prophetically, Terzi even described the launching of such a vessel, recounting a process that antedated some of the actual procedures that would be used to launch the zeppelins of the distant future. First, hold the ship down with ropes until it is ready to fly. "Then release the cords gradually and all together, so that the ship may lift itself up with the car, carrying with it many men, more or less, according to the lifting power."

Terzi well knew that many people would regard the notion of a flying ship—*a flying ship!*—as preposterous, "a mad fantasy." He therefore considered and responded to a number of more or less obvious objections to the successful operation of his device. One was that setting foot in such a contraption would be risky. Terzi readily conceded the point: "I confess that our aerial ship might run great perils, but not more than the ships on the seas are liable."

Another objection was that, once having started its ascent, there would be no way of stopping it, with the result that the ship would fly so high that "men would not be able to breathe." But the ascent *could* be stopped at any point, Terzi said, simply by letting air back into the spheres: since air has weight, it would act as ballast and thus counteract lift. The same procedure—opening the valves and letting more air in—could bring the ship safely back to earth.

A third counterargument was that the pressure of the surrounding air would flatten the spheres that enclosed the vacuum. "To this, I reply that it might so occur if the vessel were not round, but being spherical the outer air could only compress it equally on all sides, so that it would rather strengthen it than break it." Clever enough, but Terzi was in fact wrong about that: a thin, evacuated copper sphere surrounded by air would be crushed at once—essentially imploded—by normal, sea-level atmospheric pressure. After all, there would be nothing inside the sphere to resist external atmospheric forces. But the amazing thing is that, except for this defect—which is certainly not trivial—Terzi's scheme would actually work!

After answering these and other possible objections, Terzi went on to advance one of his own, "which to me seems the greatest of them all, and that is that God would never surely allow such a machine to be successful, since it would create many disturbances in the civil and political governments of mankind."

Indeed, Terzi foretold the future of the airship with an eerie, almost disturbing precision. "No city would be proof against surprise, as the ship could at any time be steered over its squares, or even over the courtyards of dwelling houses. . . . And in the case of ships that sail the seas . . . iron weights could be hurled to wreck the ships and kill their crews, or they could be set afire by fireballs and bombs; not ships alone, but houses, fortresses, and cities could be thus destroyed."

So, more than two hundred years before the first practical, working example ever existed, Francesco Lana de Terzi predicted the operation and impact of the airship in some detail and enumerated many of the craft's advantages and drawbacks. Furthermore, and essentially at a stroke, he had founded the science of aerostatics, the study of gases in equilibrium and of bodies suspended in gases. It was the basic theory underlying the operation of all balloons and airships, old or new.

A vacuum—nothing at all—was the first philosopher's stone of flight, but there were also real, physical materials with negative weight: soap bubbles, for example, as well as smoke, hot air, and fire. As an airship-lifting agent, however, hot air was far from perfect. For one thing, as its heat was lost, so was its lift. Second, the flames that produced the heated air could also set fire to the craft. These were serious drawbacks. Nevertheless, hot air was better than nothing (as it were). And as it happened, the first actual lighter-than-air craft, albeit only a small test model, used hot air as its source of lift.

The craft's inventor was, like Terzi, a little-known seventeenth-century Jesuit priest. Born in Brazil in 1685, Bartolomeu Lourenço de Gusmão was educated at Baia Seminary, where he designed a new type of water pump. He emigrated to the mother country, Portugal, at the age of twenty and joined the faculty at Coimbra University. Established in 1290, Coimbra was one of the world's oldest institutions of higher learning (antedating even Harvard!). There Gusmão put forth various schemes and designs for flying machines, of both the lighter- and heavier-than-air variety.

Gusmão is one of those blurry characters about whom no two accounts of his life, achievements, or inventions are in exact agreement. Indeed, a historian of the period described him as "one of the most shadowy and intriguing figures in aeronautical history." In one version of events, for example, Gusmão's Eureka moment came when he was washing his hands and saw a soap bubble rise up in front of him. Another version had it that the soap bubble rose in the hot air above a candle flame. Since the two accounts were not contradictory, both could be true.

It seems well established, at any rate, that in the year 1709 Gusmão wrote to the King of Portugal, John V. Not your ordinary stuffy monarch, King John was a good-looking dandy who had become sovereign in 1706, at the age of seventeen. Even at that age he was interested in both the arts and the sciences. Early in 1709, Gusmão wrote to King John requesting a patent on an aerial machine that could fly two hundred miles within twenty-four hours. According to C. H. Gibbs-Smith, a historian of early aviation, "the resulting patent does exist and is in the State Archives at Lisbon, dated 19th April, 1709. This document throws no light on the aircraft and is chiefly concerned with the political advantages which would accrue to the nation which possessed such a useful military transport."

Several other records from the period exist as well, and they describe the device in question, which was called the *Passarola* (Great Bird), as a hodgepodge of individually unlikely and jointly baffling elements that included a sail, wings, and various tubes and wires, all of it propelled by magnets inside metal spheres. It was like a prop from a comic opera, an item that Gibbs-Smith dismisses as "almost pure nonsense from start to finish."

Whatever the truth concerning the *Passarola*, Gusmão also conceived of a second, and more practical, invention that by all accounts he actually demonstrated. The exhibition occurred on the fateful date of August 8, 1709, when, according to Gibbs-Smith, "a quite definite event occurred." It was an ascent of a hot-air balloon, apparently the world's first.

An account dating from 1724, a full fifteen years after the event, states that "Gusmão's device consisted of a small bark in the form of a trough which was covered with a cloth of canvas. With various spirits, quintessences and other ingredients he put a light beneath it and let the said bark fly in the Salla das Embaixadas before His Majesty and many other persons. It rose to a small height against the wall and then came to earth and caught fire when the materials became jumbled together. In tilting downwards it set fire to some hangings and everything against which it knocked. His Majesty was good enough not to take it ill, and maintained his graciousness."

If this is indeed what happened, it would have been a kind of distant-early-warning prequel to the *Hindenburg* fire. Over the course of airship development there would be many more examples of this unfortunate but recurrent phenomenon. Gusmão is therefore one of the first pioneers of a pathological technology: an inherently dangerous machine that would perform the miracle of flight.

Before he passed from history, "dying poor and forgotten, in a hospital in Toledo [Spain]" in 1724, Bartolomeu Lourenço de Gusmão left one more design, of which he apparently produced a simple diagram. It shows a pyramid-shaped balloon, an open gondola, and a rudder. The rudder might seem out of place, inasmuch as it extends downward from the ship, as if into water. But later, working airships also had rudders, as do modern aircraft to this day. What's wrong with this design is that the craft as depicted lacks any sort of propulsion system, meaning that the rudder would have no airstream to react against.

G usmão's little *Hindenburg* prequel was merely a prototype, so small and experimental that its one and only flight took place indoors. Some seventy-five years later, lighter-than-air flight—including human flight—became a practical, working reality. It would have dismayed Ferdinand von Zeppelin that all of the technical advances were made by Frenchmen. The first and most famous were the balloons made by the Montgolfier brothers, Joseph-Michel and Jacques-Étienne, paper manufacturers from Annonay, France.

According to the legend, in November 1782 Joseph received an inspiration from the sight of ashes rising in the smoke from a wood fire. He inferred from this that hot air is lighter than the surrounding cold air, and he decided to make practical use of this difference. He soon constructed a small test balloon out of taffeta, filled it with smoke from a fire, and lo and behold, it rose into the air. The balloon reached the ceiling, as had Gusmão's model, but it did not burst into flames or set fire to the draperies. Later, on June 4, 1783, Étienne made another, bigger balloon; when launched outdoors in the Montgolfiers' hometown, it reached an altitude of 1,000 feet—or 1,500 feet or maybe even 3,000 feet. Altitude estimates at this stage of flight development were largely a matter of guesswork.

Unfortunately, the Montgolfiers' contraptions were no more steerable than the columns of smoke that they embodied, but at this point even uncontrolled flight was a considerable achievement.

The brothers went to Paris to establish themselves as lighter-than-air pioneers only to find that they had been beaten to the punch by a competitor who had harnessed for the purpose an even better lifting agent than hot air. The competitor was Jacques Alexandre César Charles, a physicist, a member of the French Academy of Sciences, and a friend of Benjamin Franklin.

It was Charles, not the Montgolfiers, who brought aerial craft to Paris. Further, his balloon was filled with a substance that was to become, in its way, the most efficient lifting agent of them all, although it exacted a great penalty for its powers. This new philosopher's stone of flight had been discovered just seventeen years earlier: it was a cold lifting gas called "inflammable air," later to be known as hydrogen.

Everything about hydrogen seems to be abnormal. It is by far the most abundant element in the universe, comprising approximately three-quarters of all matter. Nevertheless, there is next to no *free* hydrogen on earth, the reason being that since it is the lightest element, and since its molecules move very fast, any free molecules have long since escaped the earth's gravitational pull and disappeared into space, as do any hydrogen molecules that are newly released into the atmosphere today. Still, *bound* hydrogen molecules are present in water, and it's possible to separate them from their attached oxygen atoms either by means of electrolysis, which amounts to electrocuting the liquid and forcing apart its atomic constituents, or by other methods that make use of various chemical mixtures and reactions.

The man credited with the discovery of hydrogen was Henry Cavendish, himself a highly unusual specimen, and eccentric to a degree that was rare even for a Britisher. He was born in Nice because his mother, Lady Anne Grey, had traveled there to improve her health. The trip was evidently not an entire success, as she died two years later. Although he was of noble birth and quite rich, Cavendish tended to dress like a pauper. In England he attended St. Peter's College at the University of Cambridge, but took no diploma because he refused to participate in the required religious services; he seems to have been an atheist or at any rate an agnostic.

On a personal level, Cavendish found it difficult to tolerate the sea of faces around him, which is to say, those of the faculty and his fellow students. In fact, he was so paranoiacally shy that he avoided face-to-face contact, or even talking, with other people, and he was so afraid of women, including his housemaids, that he dealt with them only by means of written notes. Cavendish had a separate entrance built into his house so that he could steer clear of others as he came and went. And he kept a library four miles away from his home so that he could study there in solitary splendor and in relative peace and quiet.

All this was almost a caricature of mad-scientist behavior. Nevertheless, Cavendish was a participating Fellow of the Royal Society and communicated to its members the results of the experiments he performed in his private laboratory. In 1766, at age thirty-five, he published his first works, in the *Philosophical Transactions of the Royal Society:* "Three Papers, containing Experiments on factitious Air." (By "factitious" he meant produced by artificial means, as opposed to occurring in unaided nature.)

The most important of these experiments concerned a factitious air that Cavendish created by dissolving iron, tin, or zinc in sulfuric acid (which chemists of the time called "vitriolic acid"). "The air produced thereby is inflammable," he wrote, and for that reason he called it *inflammable air.* "When a piece of lighted paper is applied to the mouth of a bottle, containing a mixture of inflammable and common air, the air takes fire, and goes off with an explosion."

The French chemist Antoine Lavoisier later named inflammable air "hydrogen," from the Greek for "water-generated." So fascinated was Cavendish by the explosive properties of hydrogen that he performed a series of experiments that compared the result of mixing it with various different proportions of "common," or ordinary, air and then igniting the mixture.

"With 3 parts of inflammable air to 7 of common air, there was a very loud noise," he reported. "In the next three trials, though they made an explosion, yet I could not perceive any light within the bottle. In all probability, the flame spread so instantly through the bottle, and was so soon over, that it had not time to make any impression on my eye. . . . They each went off with a pretty loud noise, and without any difference in the sound that I could be sure of."

In addition, Cavendish found that inflammable air was *lighter* than ordinary air: after another series of tests and measurements, he determined that "inflammable air comes out 8760 times lighter than water, or eleven times lighter than common air." (The current value puts hydrogen at approximately fourteen times lighter than air.)

Clearly, to Henry Cavendish, the most important attribute of inflammable air was that it was explosive. This was the substance, nevertheless, upon which Count Zeppelin would later stake his hopes and his fortune, as well as the lives and safety of his future passengers.

Jacques Alexandre César Charles also pinned his hopes on hydrogen, although to a lesser degree of ambition; his first balloon, the *Globe*, would be unmanned. The *Globe* was a smallish specimen, only twelve feet in diameter and made out of alternating red and yellow strips of rubber-coated silk. Charles had produced the hydrogen by pouring a large quantity of sulfuric acid over half a ton of scrap iron in an oaken cask. The resulting chemical reaction generated a lot of heat, which he dissipated by passing the hydrogen gas first through a system of lead pipes and only then directing it into the balloon.

And so, in the late afternoon of August 27, 1783, in heavy rain, its liftoff heralded by a cannon shot, the *Globe* rose up off the Champs de Mars in Paris (later the site of the Eiffel Tower) and started floating across the city. The vision transfixed all those who saw it. According to a contemporary account, "The idea that a body leaving the earth was traveling in space was so sublime, and appeared to differ so greatly from ordinary laws, that all the spectators were overwhelmed with enthusiasm. The satisfaction was so great that ladies in the latest fashions allowed themselves to be drenched with rain to avoid losing sight of the globe for an instant."

Indeed, a later historian pronounced the sight "the most remarkable image of the eighteenth century: a balloon floating sedately above a city." It was incongruous, almost unbelievable, this event that seemed to violate the natural order of things. On the other hand, it was also the event that supposedly elicited Benjamin Franklin's response to a bystander's question "Of what use is it?" (speaking of the balloon): "Of what use is a newborn babe?"

After remaining aloft for three-quarters of an hour, the *Globe* descended into a field near the rural village of Gonesse, whose *citoyens* had not yet reached the fullness of eighteenth-century enlightenment: they regarded the vessel as an alien living monster that would have to be destroyed. The villagers therefore set upon the slowly deflating beast and attacked it with pitchforks, rocks, knives, muskets, and angry dogs, until they were sure it posed no further threat to themselves or anyone else.

A few weeks later, in Paris, the Montgolfiers sent up two balloons of the "old" type, which is to say, of the hot-air variety. The first of the two had so much initial buoyancy that during a prelaunch trial it momentarily lifted eight men off the ground until several others grabbed onto the ropes and brought it back down to earth. The second balloon, called the *Martial,* ascended on September 19, 1783, before a massive audience that included Louis XVI, Marie Antoinette, and "at least 100,000 Parisians." Inside the balloon's basket rode three animals: a sheep, a duck, and a rooster.

But the first human flight still lay ahead. Technically, this had already occurred with the brief liftoff of the eight men by the errant *montgolfier,* but that was clearly an accident, not a desired or intended effect. Also technically, another human ascent took place in October, when Étienne Montgolfier himself rose a few feet into the air on a short proving flight of an enormous tethered hot-air balloon, called by the French *un Globe Aërostatique.* However, the craft in question was more vividly remembered for its elaborate paint scheme, which made it, according to one account, "a flying Fabergé egg" and "the most spectacular decorated vehicle ever to exist." Its outer surface was adorned with the twelve signs of the Zodiac, the King's initials, multiple radiating sunbursts, and a repetitive pattern of fleur-de-lis, among other motifs.

Still, tethered ascents, whether manned or unmanned, were not true "flights" in the full and honest sense of the term. The distinction of being "the first aeronaut," of having made "the first human flight in the history of the world," in the words of aviation historian Richard Hallion, belongs to a man who was by any measure the most charming, the most amusing, and by far the craziest of any of the heroes of lighter-than-air flight, which is saying a lot. This was another Frenchman, the entertaining but reckless Jean François Pilâtre de Rozier.

Rozier, who was born in Metz in 1754, was a pharmacist, chemist, physicist, and inventor. One of his inventions was a gas mask and respirator that allowed sewer workers and others to breathe underwater. He was so enthusiastically skilled at giving public lecture-demonstrations that in 1781 in Paris, at the age of twenty-seven, Rozier started a private institute, or school, called the *Musée,* in which he gave lessons in the sciences to anyone, layman or noble, who could pay a modest fee. It was an early instance of the later activity known as "citizen science."

One of Rozier's favorite demonstrations was a circus sideshow–quality performance with hydrogen gas, a substance that he found fascinating. Rozier would inhale a quantity of the gas, blow it out through a short pipe, and then ignite it, producing unusual visual and auditory effects and "a very curious jet of flame." On another occasion, "the intrepid physicist . . . mixed the very pure gas with one-ninth its volume of air and then inhaled this mixture in his usual fashion. But when he tried to ignite it there resulted such a terrible explosion that it was feared his teeth had been carried away."

So in all fairness, who better than he to become the first man of flight?

The momentous event finally came to pass on Friday, November 21, 1783, at 1:54 p.m., when Rozier, accompanied by the marquis François Laurent le Vieux d'Arlandes, departed this earth, free of all tethers and restraints, and lifted off from the Bois de Boulogne in the very same *Globe Aérostatique* in which Étienne had already made his short tethered hop. Benjamin Franklin was again in the audience. Another observer, the Duc de Polignac, reported that the spectators (including himself) "experienced a feeling of fear mingled with admiration. Soon the aerial navigators were lost from view, but the machine, floating on the horizon and displaying a most beautiful shape, climbed to at least 3,000 feet at which height it was still visible; it crossed the Seine below the gate of la Conférence and, passing between the École Militaire and the Hôtel des Invalides, it was borne to a position where it could be seen by all Paris."

But Rozier's final act and *coup de grâce* was yet to come: an attempt at crossing the English Channel aboard his newest, most novel, and by far very worst invention: the combination hot-air-and-hydrogen balloon. This consisted of a hydrogen-gas sphere set atop a cylindrical hot-air balloon, the two held together by netting and ropes, producing an ensemble that looked like a golf ball upon a tee. The idea behind the two-part device, which Rozier called an *Aéro-Montgolfière,* was to combine the brief but great lifting capacity of hot air with the relatively long life span of hydrogen. The hot air was to be provided by the heat from a small portable stove burning straw and other materials, which meant placing an open flame beneath a substantial quantity of explosive gas, an arrangement that a critic said was like "putting fire beside powder." Indeed, this was a veritable masterpiece of pathological technology, a paradigm case of "really asking for trouble." But Rozier could not be dissuaded from his *idée fixe*—flying a crossbreed hot-air-and-hydrogen combination vehicle across the water to England.

He and fellow crew member Pierre Romain lifted off from the French coast a little after 7:00 a.m. on June 15, 1785. They reached a height of a thousand feet or so before the winds blew them back to the shoreline. And as if this were an interactive video, the story then unfolded along two separate narrative lines, each of which was accompanied at the time by an engraved illustration that depicted each of the two different endings.

In one account, the craft plummeted to earth and killed both men. Whether miraculously or suspiciously, there was no conflagration.

In the other dénouement, both aeronauts died when "a violet flame appeared at the top of the balloon, then spread over the whole globe, and enveloped the Montgolfière and the voyagers." In this case, the crash of the balloon and its destruction by flames would make the finale into *Hindenburg* prequel number 2. Either way, it was the world's first aerial disaster.

Despite its flammability and its tendency to explode at the least provocation, hydrogen gas soon became the airship lifting agent of choice.

Chapter 3

THE FLYING BOMB

One hundred years after Pilâtre de Rozier and Pierre Romain crashed to the earth and died aboard their unique, custom-made, hot-air-and-hydrogen composite balloon, Reynard and Krebs in 1884 were successfully flying their electric-powered dirigible *La France*. This was a development that Ferdinand von Zeppelin viewed with alarm. Such a craft, he feared, could conceivably pose a military threat to Germany, one that should therefore be counteracted by an even bigger and better airship. Six years later, in 1890, Zeppelin was out of the army and quite unexpectedly had both the time and opportunity, as well as the motivation, to foster German airship development and to bring the country into aerial parity with the French. And so he did what he could to invent an airborne vehicle of his own.

It had been back in 1874 when Zeppelin had first cobbled together a few assorted "thoughts about an airship." These described a large, aerodynamically controlled craft whose interior was divided into several individual gas cells. Zeppelin was aware that as a former cavalry officer with no expertise in aeronautical engineering, he would need some help in converting his abstract ideas into practical realities. And so, in May 1892, Zeppelin hired Theodor Kober, a man who had previously worked as an aeronautical engineer for the Riedinger Balloon Factory in Augsburg. The two of them came up with a surprising new design that was quite remarkable in its way. Their proposed airship was in fact a train—specifically, an "aerial express train." This was basically three individual airships attached together in sequence, each performing a different function: a pulling or tractor unit (locomotive) powered by two engines, followed by a passenger car (coach), and ending up with a cargo section (caboose). Just as some early airplane designs made use of birdlike (or batlike) flapping wings, so too the aerial express train seems to have been the product of an overly literal transference of the structural features of a terrestrial model—in this case railroad cars—to airborne devices. On the other hand, the railroads of Zeppelin's era were undeniably successful at fast and efficient mass transit, so why *not* move the railroad cars off the ground and into the sky, where they could fly through the air with the freedom of birds?

In any case, Zeppelin thought so much of this novel design that in 1893 he wrote a letter to General Alfred von Schlieffen, Chief of the General Staff (and later the originator of the "Schlieffen Plan" for victory in World War I), in which he extolled the virtues of his invention and explained how it would function as a vehicle of war: "Safe and fast-flying air trains will confer on the Army many advantages: for example, reconnaissance over hundreds of miles of enemy territory in a few hours; assured supplies for all troops from the nearest depots every day without the delays of road transport; transfer of important officers and important information from one army to the other; bombardment of enemy fortresses or troop concentrations with projectiles, etc." In every way, this was the fulfillment of Francesco Lana de Terzi's nightmare vision for his "aerial ships."

The Count never said exactly what part Kober might have played in coming up with or refining this three-part invention; in fact, Zeppelin somewhat equivocally proclaimed the concept as his own, while also giving an unspecified amount of credit to Kober: "The machine is my own invention only so far as concerns its fundamentals," he wrote. "The whole merit for its completion, which can challenge comparison with such modern engineering problems as the Channel tunnel, the Eiffel Tower, etc., belongs to my engineer, Herr Kober."

Whatever the proper division of credit (or blame) for these somewhat strange ideas, Zeppelin and Kober had in fact worked out their plan in substantial detail. Each separate car of the airship train would have the same basic architecture of all the zeppelins ever to be produced in the future: a rigid internal framework consisting of a series of rings connected to horizontal girders, with individual gas cells occupying the open spaces between the rings. The rings themselves would be stabilized and braced by cables that crisscrossed each ring like the spokes of a bicycle wheel.

The use of several individual gas cells to contain the hydrogen was an innovation. Apparently, this was for redundancy: if one or more cells leaked or failed, the others would suffice to keep the ship airborne. In addition, if the cells were made of a gastight material, then (in theory) no hydrogen would leak from the craft. This was another design element that was slightly reminiscent of Terzi's "aerial ship," whose multiple spheres of nothingness would be gathered together to suspend the device in midair. Zeppelin placed the cells in a line, put them inside a metal cage, and enveloped the whole structure in a fabric covering of "silk or similar material."

But for all its novel features, the German military took a dim view of Zeppelin's "aerial express train." In 1894, after several airship specialists, including Rudolf von Tschudi, commander of the Prussian Airship Battalion (a balloon brigade), Professor Heinrich Müller-Breslau of the Technical College at Charlottenburg, together with several other experts, had examined Zeppelin's plans, at his invitation, the group reported to General von Schlieffen that "we must advise the War Ministry not to involve itself with the execution of the project of Major-General Graf Zeppelin." The structure as designed was too flimsy, they said, and the engines of the "locomotive" section were too feeble to drag the rest of the train, plus cargo, through the air. They added, nevertheless, that "even though the concepts at the basis of the invention are not new, in the execution many notable concepts were found and one had to credit the project with a certain originality." That much was true.

Even after this rebuff, Zeppelin applied for a patent on his (and Kober's?) invention, and on August 31, 1895, Graf Ferdinand von Zeppelin (alone, without Kober's name on the document) was granted Imperial German patent number 98580, for a "Controllable air train with several lifting bodies in series," an apparatus that the patent office categorized under the heading "Class 77," or "Sport," although what sport an aerial express train might be used for it did not say.

Later, in 1897, Zeppelin also applied for—and in 1899 was granted—an American patent (number 621,195) on his design for the airship, which he now called a "Navigable Balloon." The patent application made no mention of the craft's dimensions or gas volume, but it did say that directional control was to be effected by means of tiny rudders "at the front or the rear part of the balloon," whereas the inclination of the craft was to be achieved by means of sliding weights slung underneath each of the three cars.

When news of the patent became public, the overall concept was lampooned in predictable ways by the American press. "The locomotive and trailers are all to be bologna sausage shaped," said the *News and Courier* of Charleston, South Carolina, on November 11, 1900. This was accurate enough: the rounded front and back ends were so similar that it was practically impossible to tell which was which. The train itself, furthermore, "will have the appearance of a string of giant sausages." Still, the patent application also explained that "the space between each two balloons is closed by means of an extensible cover, which lies over the cylindrical shells of the two adjacent balloons, so that the wind cannot obtain a hold in the intermediate space." And so, "with these coverings in place," the *News and Courier* concluded, "the connected train will appear like a gigantic elongated worm."

A gigantic elongated worm! Poor Zeppelin. Despite his good intentions, hard work, and tireless self-promotion, in the end his "aerial express train" never got rolling.

In the meantime, someone else had invented the rigid airship—at least the first one that actually flew. This was the Croatian businessman David Schwarz.

It may or may not be an accident of history that in the storied annals of airship conceptualization and development one will find an oversupply of offbeat, iconoclastic, eccentric, and hilarious characters, and possibly even one or more crackpots. David Schwarz, whose "life has been hidden in mystery," according to one historian, was no exception to the rule.

Like Zeppelin, Schwarz was a layman, not a scientist. He was born in 1850 (or 1852—another mystery) in the city of Keszthely, located on the western shore of Lake Balaton, about a hundred miles southeast of Budapest. He was largely self-taught and at an early age became a timber merchant. He seems to have had a desire to fly from boyhood, and at some point he designed an airship. It had a cylindrical body, a cone-shaped nose, and a slightly rounded back end. It looked like a pencil stub, perhaps, or a bullet or an artillery shell. The craft had three principal features in common with the airships that Zeppelin would go on to design and build: it had a rigid framework consisting of horizontal girders and a series of vertical, wire-braced rings; it would be made out of aluminum; and it would use hydrogen for lift.

There was also one big difference: whereas the hydrogen inside Zeppelin's ships would be enclosed in a series of individually gastight bags, the hydrogen in Schwarz's airship would drift around freely inside the ship as a whole and be held in only by its aluminum outer skin. Schwarz's craft, in other words, had a metal exoskeleton that itself was supposed to be gastight. In two respects, this was a naive idea. For one, it would be very difficult to make the riveted cladding of a large airship impervious to the seepage of extremely small-molecule hydrogen gas. But if it actually was gastight, then the ship would be vulnerable to the expansion of the gas as it rose to a higher altitude, and to its contraction during descent. So the device, if actually built to be leakproof, would in effect be a hermetically sealed tin can that could explode or implode with differences in temperature or external atmospheric pressure.

Knowing that this was too big a project for him to finance himself, Schwarz had originally asked the Austro-Hungarian War Ministry to underwrite construction of the craft, but had been turned down. He then offered his device to the Russians, who, unlikely as it was, accepted. Separately, Schwarz also persuaded the German aluminum manufacturer Carl Berg to commit himself to providing the metal for building the flying machine. On August 23, 1892,

Schwarz and Berg signed a contract whereby they would jointly build and fly the world's first navigable all-metal, rigid airship, in Russia. By November, Berg was shipping the parts, including 100 sheets of aluminum cladding, plus a grand total of 269,000 rivets, collectively weighing 100 pounds, to an airfield in St. Petersburg, where the Russians were erecting a construction shed.

Schwarz and Berg ended up building two airships. The first (*Schwarz I*) was designed to be 68 feet long and 39 feet in diameter, with a conical nose ("beak") and a rounded stern section, the whole housing some 91,800 cubic feet of hydrogen. But the ship as actually built was larger in every dimension and in the end held 151,800 cubic feet of the lifting gas. In addition, Schwarz had miscalculated the length of the propeller brackets, and so he had to make indentations in the aluminum covering to allow the propellers to turn without slicing through the hull. Rumors had it that the device would be powered by a steam engine . . . or maybe it was by a 10-horsepower Daimler gasoline engine.

Inflation began on August 19, 1894. Schwarz had boasted that, once filled, the hydrogen inside would last for more than a month, but as it happened, half of the hydrogen leaked out of the ship overnight. The workers tried inflating it again on August 27, and this time the process was going along so smoothly and efficiently that Schwarz predicted it would be completed within an hour.

And at that auspicious point, everyone broke for lunch. During which time the ship imploded in upon itself and collapsed to the ground in a heap.

David Schwarz left St. Petersburg and returned to Zagreb, where he was then living. One of the Russian engineers who had worked on the project, a man by the name of Kowanko, had said of the craft: "In the light of present technology, the directional control of the airship is inconceivable, and it is unsuitable in all its features and design."

But such negativity was nothing to a dedicated, visionary prophet such as David Schwarz, nor to his partner, the aluminum magnate Carl Berg. While the remains of their first ship collected dust in its hangar, Berg, in an evident tour de force of high-pressure salesmanship, managed to win a contract to build a follow-on craft, the *Schwarz II,* for the German Imperial Government. Unaccountably, the Germans were more impressed by Schwarz's plans for a flying aluminum artillery shell than by Zeppelin's design for an aerial express train.

Construction of *Schwarz II* began in the Prussian Airship Battalion balloon hangar on Tempelhof Field in Berlin during the summer of 1895. The new ship was to be 156 feet long, and with the same conical nose and rounded stern sections as before, but this time it would have an elliptical cross-section that was 39 feet wide by 49 feet high. Gas volume was 130,650 cubic feet, and the exterior aluminum sheeting, riveted to the hull, was once again relied upon to contain the hydrogen gas that filled the interior. The ship had pusher propellers on each side, plus another one mounted horizontally on the bottom. It had no aerodynamic control surfaces; right-left steering was to be accomplished by the use of differential thrust from the port and starboard engines, while ascents and descents were to be managed by the horizontally mounted airscrew.

It took a year to build *Schwarz II.* Inflation began in August 1896, but the gas leaked out just as it had with the earlier model. There was another attempt on October 9, and that too proved unsatisfactory: this time, because the hydrogen was not of sufficient purity and did not provide enough lift, the ship would not rise from the ground. Yet another try at filling the craft, with purer-quality hydrogen, was scheduled for January 13, 1897.

But on that very day, on a trip to Vienna, David Schwarz died, at the age of forty-four. Theories abound as to the cause of death, with explanations ranging from a heart attack due to all the excitement, to a claim that he collapsed "after receiving a telegram informing him of the government's interest in his studies," to food poisoning. Incontestably, he died of death. He was buried in the Jewish section of a cemetery in Vienna four days later, and his gravestone bore the fittingly obscure inscription: THE EARTH WAS FOR HIM LIGHTER THAN LIFE ITSELF.

Whatever that meant, Schwarz had left behind no instructions about who was to pilot his gleaming silver creation, nor precisely what procedures would be followed during the flight test. This was unfortunate inasmuch as nobody in Germany had yet flown a powered airship, meaning that both the test pilot and the craft itself would be venturing into the unknown.

But Schwarz did leave behind a wife, Melanie, who now teamed up with Berg to get the craft into the air. That finally happened on November 3, 1897, after the ship was fast-filled with hydrogen in a speedy two hours and forty minutes, and then, before the gas had a chance to leak out, a Prussian Airship Battalion mechanic by the name of Ernst Jägels (known to be "cold-blooded and resolute," but who had never before flown even so much as a balloon), climbed into the gondola.

The craft looked like an enormous aluminum-foil rocket ship, its thin metal covering wrinkling and rippling overhead in response to the various tensions, stresses, strains, twists, and pulls placed upon it by its internal hydrogen, plus the weight of its engines, gears, drive shafts, and other cogs and wheels, as well as by the gusty winds.

The plan was that this would be a safe, easy, tethered ascent, but the ship when released turned out to have so much excess lift that it immediately broke free of its ground crew and rose to a height of about 400 feet. (Such behavior would soon become a recognized part of an airship's bag of tricks.) At that point, the port propeller's drive belt slipped off its drive pulley, and the ship swung broadside to the wind. The craft then continued its ascent up to 1,700 feet, at which altitude the right-side drive belt also left its shaft behind.

The airship was now officially out of control and was essentially a free balloon, a captive of random air currents, but at least it was flying. Not for long, however, as Jägels, clearly panicked, opened the gas release valve, whereupon the ship descended, hit the ground, rolled over like a dog, and deflated. Jägels managed to jump clear without injury.

Short as its maiden (and only) flight was, the device nevertheless established David Schwarz, not Count Ferdinand von Zeppelin, as having created the first rigid airship that actually flew. But to the extent that Schwarz is referred to at all in biographies of the Count or in histories of the airship, he is mentioned only in passing, demoted to a footnote, or otherwise marginalized. There are some reasons for this. For one thing, although Schwarz's craft did in fact fly, his device was such a structural, dynamic, and mechanical mess that it was effectively uncontrollable practically from the moment it left the ground. (Still, one of Zeppelin's own airships crashed even before fully leaving the hangar, while another one burned to a crisp while still *inside* the hangar.) Second, Schwarz and Berg worked in great secrecy, which posed a challenge to later historians of flight, whereas Count Zeppelin routinely broadcast his ideas to all who would listen (and even to many who wouldn't). Third, there was the fact that Schwarz did not live to see his own craft take to the air. And finally, there was the legend that after the ship's one and only ascent Schwarz's widow accused Zeppelin of having stolen and used some of her husband's ideas, and that in reprisal she tried to extort money from the Count, thereby adding an element of scandal to a story in which everything that could possibly have gone wrong seemed to have done so already.

Hugo Eckener, in his biography of Zeppelin, provides a different account of what happened. After the Schwarz ship crashed, "Count Zeppelin negotiated with Herr Berg's firm for the purchase of the aluminum for his own ship. The firm, however, was under contract to supply aluminum for airships exclusively to the Schwarz undertaking. It had to obtain release from this contract by arrangement with Schwarz's heirs before it could deliver aluminum to Count Zeppelin. That is the origin of the legend."

It was also the start of a grand tradition in the airship business— the reuse of aluminum from a crashed airship in the next new model. In time, zeppelin builders would become past masters at the art of metal recycling.

Along with hydrogen, aluminum was one of the two chemical elements that made airships possible. When the *Schwarz II* flew in 1897, however, the aluminum industry was only about ten years old. And although it was the third-most-abundant element in the earth's crust, comprising 8 percent of the solid surface of the earth by weight, aluminum did not naturally exist in its pure metallic form, as did, for example, gold, silver, and copper. Rather, aluminum appeared only in combination with other substances from which it had to be forcibly separated.

It was for this reason that aluminum was not isolated as an element until 1825. Thirty years after the Danish physicist Hans Christian Oersted isolated the metal, it was still an extremely rare commodity, and in 1855, when the French chemist Henri Étienne Sainte-Claire Deville produced it in small, impure amounts, it was literally more expensive than gold. Indeed, one of the first artifacts to be made from aluminum, in 1856, was a baby rattle fashioned for the son of Napoleon III (who had financed Deville's work on the metal).

Aluminum became an industrial commodity through the independent discoveries of two men who seemed to be mystically linked by some sort of spooky action-at-a-distance phenomenon, like quantum entanglement: Paul Héroult in France and Charles Hall in the United States. The least of it was that both of their last names began with an H. Héroult was born eight months before Hall in 1863, and died eight months before him in 1914. In between, they discovered the same process for producing aluminum, weeks apart in the same year, 1886, when both of them were the same age, twenty-two.

Hall and Héroult each separately found that high-purity aluminum could be produced in batch quantities by electrolysis, which was also the process by which oxygen could be separated from hydrogen. The key step was dissolving aluminum oxide in a molten mineral called cryolite to lower the oxide's melting point. Then, by passing an electric current between carbon electrodes, molten aluminum was deposited at the bottom of the electrolytic cell, from where it could be tapped off.

By 1888, two years after the discovery of the Hall-Héroult process (which is still in use today), the price of aluminum dropped from a record $500 per kilogram to less than $4. Low-cost electricity, in turn, was made possible by the roughly simultaneous development of the dynamo. Because of its strength, lightness, and cheapness, aluminum soon became the obvious material of choice for building airships, starting with David Schwarz and continuing on through, and well beyond, the reign of Count Ferdinand von Zeppelin.

By the time the *Schwarz II* flew in 1897, Zeppelin had raised himself from the lowly status of a dismissed and humiliated cavalry officer to the exalted rank of a prophet, a messiah, the Prometheus of lighter-than-air flying machines. This extreme career makeover would in the end accomplish a number of objectives. First and foremost, as the pioneer of airship invention, development, and production, he would, if successful, rescue Germany from its position as a second-rate power in the realm of the air. Second, he would bequeath to the mother country a grand new, practically invincible (as he thought) weapon of war. And in so doing, he would restore his good name and reputation and acquit himself in the eyes of his peers. For Zeppelin, this was clearly a win-win-win situation.

But to do all this he needed two things: he needed credibility, and he needed money. The first he got from an unlikely source, the Union of German Engineers, a professional group to which he applied in 1896 for scientific and technical support, outlining a plan to build what was essentially the initial, locomotive section of his "aerial express train." Surprisingly enough, this august body championed his cause, saying, among many other things, "Theoretically, it is agreed that natural laws present no obstacle and that existing technical resources are sufficient to meet the static and dynamic requirements of airship construction. In the opinion of distinguished physicists and engineers the difficulties and objections are no greater than those which faced existing techniques before the days of modern shipbuilding and railways. The object of these endeavors is: safe transport in the air, independent of all kinds of roads, at speeds hitherto unattained."

As for money, Zeppelin canvassed what were, to him, all the likely sources, including the King of Württemberg and even the Emperor; for his trouble, he wound up with only 6,000 German

marks, a pittance. He tried to raise 800,000 marks by public sub-
scription, but came away with only a few thousand, largely from
some friends of his. He was so totally serious about airships, however, that he was willing to use his own money if necessary—as
well as his wife's—to get them designed, built, and into the air.
Finally, in May 1898, Zeppelin formed a "Joint Stock Company for
the Promotion of Airship Travel." The company, which consisted
of Zeppelin himself, Carl Berg, and other industrialists, duly put
together a total of 800,000 marks, of which Zeppelin himself con-
tributed anywhere from 300,000 to 400,000, Berg 100,000, and
others smaller amounts. And with that, he and they were now very
much in the airship business.

Item one on the corporate agenda was to build the company's
first flying machine, which would be known as LZ 1, for *Luftschiff
Zeppelin* 1 (Zeppelin Airship 1). With his cosmic self-assurance
and inner certainty about everything, Zeppelin was not one for
starting out small with scale models, mock-ups, prototypes, or the
like. No, he would build big from the beginning, for, as he had
thought from the outset, to be useful these things had to be huge.
This was not simply a matter of grandiosity on Zeppelin's part,
although there was nothing more grandiose than a zeppelin (or
for that matter, Zeppelin himself). There was in fact a good scientific and engineering reason for building extremely large airships,
and this was contained in a simple cube law: while the linear dimensions of an airship rise as the square, its volume rises as the
cube. So, for example, doubling the linear dimensions of an airship increased its volume eightfold, while enlarging it ten times
increased its volume by a thousand.

The LZ 1, therefore, would be 420 feet long—longer than a
football field, a distance not quite half the height of the Eiffel
Tower, and three-quarters the height of the Washington Monu-

ment. With a diameter of 39 feet, the ship, when fully inflated, would hold 399,000 cubic feet of hydrogen. But even with the hydrogen parceled out into 17 individual gas cells, there was still no getting around the fact that the ship was essentially an extremely large flying bomb.

Needing a construction shed to build such a vast thing, Zeppelin chose the town of Manzell, near Friedrichshafen on the shore of Lake Constance, for the site. After all, that was where he had grown up, and he well knew the winds and weather of the area, which was an important consideration inasmuch as Zeppelin had, of course, decided that he himself and nobody else would pilot the craft's first flights.

The construction shed, an immense, three-story-tall wooden hangar, would not be built on land, but rather on the lake, floating on pontoons so that it could be turned into the wind for launches and recoveries. The airship itself could float on its two open-air, metal, watertight gondolas, which would also hold the engines and crew, Zeppelin's notion being that it would be easier and safer to take off from and alight upon the water than on land. A thin and rickety gangway stretched between the two gondolas.

No sooner was the shed complete than it was damaged in a storm. The repairs to it alone amounted to some 100,000 marks.

Construction of the ship itself began on June 17, 1898, and took almost two years to complete. The craft that finally emerged from the floating hangar had the classic, bilaterally symmetrical, cigar-shaped profile of many early zeppelins, with the front end visually indistinguishable from the back.

This first great monster, with an empty weight of 9,100 pounds, would be propelled through the air ("at speeds hitherto unattained") by two four-cylinder engines, each of which generated only 14 horsepower, far less than that of the average motorcycle, and in fact comparable to that of a modern riding lawn mower. There were no fixed, stabilizing fins at the aft end (nor anywhere else), such as the feathers on an arrow. Lateral control would supposedly be provided by two tiny rudders, top and bottom, at the nose, plus two more, one behind the other, at the rear.

The ship's pitch control mechanism was the same as that specified for the "aerial express train." Slung beneath the craft like a pendant on a necklace was a 220-pound lead weight that could be winched back and forth along a cable that dangled precipitously toward the ground from positions near the front and aft gondolas beneath the ship. Bringing the weight forward would pull the nose down, while bringing it rearward would have the opposite effect. When near or on the surface, the entire apparatus could be retracted. By any standard, this was a primitive, kludgy arrangement, practically guaranteed to fail.

The maiden flight of the LZ 1 took place in the evening of July 2, 1900, with the sixty-two-year-old Count Zeppelin in command, at least in theory. The impending launch had drawn a number of spectators to Lake Constance, but what that number actually was has never been determined precisely, or even approximately. Various different accounts speak of: "only Zeppelin's personal friends, members of the press and interested men and women from the neighborhood," "onlookers," "crowds," "large crowds," "thousands," "thousands and thousands of people," and "a festive crowd of some twelve thousand Germans, Austrians, and Swiss," all of them there to witness what promised to be an unrepeatable and transcendental Wagnerian moment in the history of the world.

And then, out on the lake about 600 yards from the shore, a small launch pulled the LZ 1 from its shed. The airship rested calmly on the water. Zeppelin, identifiable by a gray walrus mustache and white yachting cap, proceeded to say a prayer or two (but apparently not enough). He and four others then boarded their respective gondolas. Hugo Kübler, Zeppelin's technical manager during construction, now excused himself from stepping aboard his own creation, on the pretext that Zeppelin had not arranged for accident insurance. The Count never forgave him for this.

At 8:03 p.m., members of the handling crew released their grip on the gondolas. But, as if establishing a precedent for future ground crew mishaps, the forward team let go before the aft crew members did, and in response the nose of the ship tilted steeply upward. Then the whole thing rose into the air and drifted away.

To correct for the nose-up attitude, the onboard crew winched the movable 220-pound lead weight forward. This action returned the ship to a horizontal position, but at that point the winching mechanism jammed, leaving the weight stuck where it was. The craft was now nose-heavy and poised for a dive into the lake. To avert this the crew dropped ballast from the forward gondola. Then, almost as in a replay of the flight of the *Schwarz II,* an engine stopped running when its crankshaft broke. Finally, one of the tiny rudders stopped responding to control inputs (not that this actually made much difference aerodynamically).

After eighteen minutes of constant and unbroken fiasco, and not knowing what else might go wrong next, Zeppelin ordered that the craft be landed back on the lake. After alighting, the wind, which was stronger than the zeppelin's tow boat, dragged the airship toward the shore, which the ship struck, slightly damaging the hull. The craft had traveled a total distance of three miles, at the average speed of eight miles per hour.

The reporter covering the event for the *Frankfurter Zeitung* wrote that the flight "proved conclusively that a dirigible balloon is of practically no value." The War Ministry representative who was there for the show said that the LZ 1 was "suitable for neither military nor for non-military purposes." Which didn't leave a lot of purposes.

Further, the "rigid" airship proved to be not so rigid after all: the frame had buckled slightly during flight, the middle bending upward with the two ends sagging down, a phenomenon known as "hogging." Obviously, some alterations were in order. To make the changes, Zeppelin's workers suspended the ship from hooks on the hangar roof and then deflated it. With the hydrogen gone, the ship's weight was enough to break some of the hooks, and the center section dropped to the floor with a plop, more or less as if this were a *Looney Tunes* animated comedy film starring Daffy Duck.

Zeppelin, together with his engineers and workers, made a bare minimum of alterations to the structure of the LZ 1. They removed the cable from which the sliding weight was slung and replaced it with a walkway between the gondolas. The weight, newly increased to 330 pounds, now resided in a little wagon that traveled back and forth on rails across the walkway, like a playground trolley car. They added a horizontal elevator at the nose and repositioned the rear rudders. That was the extent of the changes.

Hugo Eckener would be in the audience for the second flight of the LZ 1. He and his wife Johanna had by this time moved to Friedrichshafen, into a house on Meersburger Strasse, near the water. Eckener had become a stringer for the *Frankfurter Zeitung,* which had assigned him to report on these breaking news events.

On October 17, 1900, Eckener watched the flight of the revamped craft. From a vantage point on a bluff at the lakeshore, he observed the airship through a telescope. He saw the ship rise into the sky . . . and not go anywhere. Instead, it milled around above the lake for an hour and a half. It made a series of small turns but otherwise remained virtually at the same height, and at the same place, for the entire time. What Eckener had at first interpreted as deliberate turns were in fact so slow and slight that he wondered if these apparent "maneuvers" weren't just the craft's being pushed around by local air currents over the lake.

"There was no question of the airship flying for any appreciable distance or hovering at various altitudes," he wrote for the newspaper. "One had the feeling that they were very happy to balance up there so nicely, and indeed the fine equilibrium of the airship was the most successful aspect of the whole affair. But under what circumstances were the modest results, which I have described, achieved? Under the best possible conditions—an almost complete calm!"

A third and final flight took place a week later, on October 24, this time as a command performance before the King and Queen of Württemberg. On this special occasion the ship raced to its maximum speed, which turned out to be about seventeen miles per hour. The flight lasted for all of twenty-three minutes, and that was that.

Eckener wrote of the ship's performance: "It could only be said definitely that both its stability and steering-gear needed improving and that the speed of 7–7.5 meters per second [17 miles per hour], measured more or less correctly, fell below expectations."

At the end of it all, a member of the Union of German Engineers, the group that had recommended construction of the thing to begin with, said of the craft: "The monster will never rise again."

And in fact it did not. Zeppelin had the floating hangar dragged to and placed upon the shore, he fired most of his personnel, and he had the LZ 1 dismantled and its metal sold for scrap. The Joint Stock Company, its funds exhausted, and with no further prospects, was dissolved.

Count Zeppelin was normally the personification of optimism and ebullience, but even he was somewhat chastened by the experience. "I shall not go on building anymore," he said.

B ut he did. It took him a couple of years to recover from this
first debacle, but by 1902 the Count was once again looking
around for money to build an airship. Seemingly obsessed with
the idea, he published an "Emergency Appeal to Save the Airship"
in the German weekly newspaper *Die Woche,* to no avail. He sent
out letters to various captains of industry, asking for contributions,
and he even enclosed stamped, self-addressed envelopes for re-
plies, but this direct-mail approach netted him only 8,000 marks.
Finally, in 1904, the year after the first flight of the Wright brothers, Zeppelin's longtime supporter, the King of Württemberg,
authorized a state lottery to save the zeppelin. This was fairly successful, bringing in 124,000 marks, but still nowhere near enough
to build the next airship. Himself out of money, having already
lost 300,000 to 400,000 marks in the failed Joint Stock Company,
Zeppelin now mortgaged his wife's estate in Latvia for the sum of
400,000 marks.

The decision to build the second ship seems to have been motivated in part by the advice of his chief engineer, Ludwig Dürr.
Zeppelin had hired him in 1899, when Dürr was fresh out of the
Royal School of Engineering in Stuttgart, to help him design and
build the LZ 1. Dürr would go on to play a major role in designing
every subsequent airship, including the LZ 129 *Hindenburg.* He
was by all accounts a notably reserved, even taciturn individual,
and hard to get along with. His single redeeming feature was that
he was a born engineer and had an intuitive understanding of how
to make things work, even airships. Up to a point.

In the spring of 1905, Zeppelin and Dürr were ready to begin
the construction of a new craft, the LZ 2. Dürr had made several
changes to the previous ship's design. For one thing, he got rid of
the amateurish sliding-weight system. He did not, unfortunately,
replace it with control surfaces of equal or greater effectiveness.

On the other hand, where the LZ 1 was constructed with I-beam girders, which were easily bent, the new ship would be made with beams that had a triangular cross-section, like that of an elongated, open prism, a shape that was resistant to bending forces in all directions. Finally, engine technology had improved in the interim, and the new ship would be driven by two 85-horsepower engines, which would provide adequate power—at least for as long as they kept running.

But in the end, none of it made any difference. On the date of its planned maiden flight, November 30, 1905, one of two things happened. Either, according to one account (by historian Douglas Robinson), "a knot in the tow rope caused it to jam when released, pulling the airship's bow into the water," or, according to another (by Hugo Eckener), "the machine failed to rise and was carried along the surface of the water far out into the lake. It was with difficulty recaptured and brought back to the hangar comparatively undamaged." In neither version of events did the ship get airborne.

A few weeks later, on January 17, 1906, Team Zeppelin tried again. This time the ship rose to an altitude of 1,500 feet. But it thereupon developed an erratic longitudinal oscillation, pitching alternately nose-up, then nose-down. Zeppelin, who was piloting the craft, later described what happened next: "While I was feeling my way, a few small mechanical disturbances set in which caused the temporary stoppage of first one motor and then the other. The steering system, too, stopped functioning altogether. Under these circumstances I was forced to land and this landing—also because of my lack of experience—had to be performed while none of the machinery was functioning."

And so a crew member simply threw an anchor overboard, after the manner of an old salt of the seas. The anchor hooked something on the ground, and this dragged the ship out of the sky. The craft ran into a stand of trees, which caused some damage. But that was the least of it.

During the night the wind increased to gale force, ripped the LZ 2 from its tie-down ropes, and effectively destroyed it. Zeppelin ordered that this ship too be torn apart and the metal sold for scrap.

Hugo Eckener traveled to the site, which was near the town of Kisslegg. "The airship lay in front of us like the skeleton of a giant whale," he said. "Men worked with axes and saws, cutting it to pieces."

And once more Zeppelin vowed: "I shall build no more airships."

Chapter 4

THE DELIRIUM

I n the light of events hitherto, Count Ferdinand von Zeppelin could not be called a success by any stretch of the imagination. Both of his first two airships had ended up as scrap heaps not long after their maiden flights. Still, the Count was not one to take criticism lightly or gracefully, even justified or constructive criticism. He was irked by the newspaper accounts of his airborne misadventures, particularly the stories that appeared in the *Frankfurter Zeitung*. About the LZ 2, for example, the author, "Dr. E," had written: "As soon as the wind or the speed of the airship brought strong air pressure to bear on the hull, distortions must have occurred, preventing the long, rigid transmission systems from the gondola to the propellers and the steering from functioning. In other words, the form of the aircraft is ill-adapted," and so on.

Zeppelin soon discovered that "Dr. E" was Hugo Eckener, who happened to be living in Friedrichshafen, not far from where Zeppelin himself had rooms in the Deutsches Haus hotel. So one day Zeppelin stopped in to see him.

It was late afternoon, but the Count was dressed in a morning coat, top hat, walking stick, and yellow gloves, as if outfitted for a night at the opera. Zeppelin came right to the point and said he wanted to correct a few of Eckener's misinterpretations of events. For one thing, Eckener had said in print that the LZ 2's engine failures were due to distortions in the power transmission system. "The real cause of the engines failing was undoubtedly the carburetors," the Count said. "They are either getting no gas at all, or they are flooded."

Zeppelin conceded that the craft was longitudinally unstable. But he already had a fix in mind for that, a modification that had been proposed to him by a Professor Hergesell of the University of Strasbourg: "He has suggested that we build small, rigid, stabilizing surfaces into the stern of the airship. Like feathers on an arrow."

A couple of days later, at Zeppelin's invitation, Eckener had dinner with the Count at his hotel. Zeppelin poured out his life story, telling Eckener about his trip to America, about his balloon ascent in St. Paul, his dismissal from the army, the continual rejection, ridicule, and criticism of his ideas, and the loss of much of his personal fortune in building his first airship, the LZ 1. Nevertheless, he spoke as if he intended to go on building even more airships, no matter what.

Eckener, for his part, ventured the opinion that in that case what Zeppelin needed was a publicist, a sort of airship impresario. And almost before he knew what was happening, Eckener had volunteered himself for the job. Which is how it came about that Eckener got Zeppelin's essay, "The Truth About My Airships," published in the German press in February 1906.

Shortly afterward, in May, just months after the LZ 2 episode, Zeppelin and Dürr were back in business. They were now working on the LZ 3, thanks to the proceeds from another state lottery plus some funds that had come directly from the Kaiser himself. In the interim they had made an important advance in the design of the craft: on the basis of Professor Hergesell's recommendation, as well as experiments that Dürr had been performing with a homemade wind tunnel, the LZ 3 would be equipped with "tail feathers," also known as an *empennage,* a French term stemming from the verb *empenner,* which means "to feather an arrow." The new ship would have tail feathers consisting of a pair of large horizontal stabilizers on each side of the rear of the craft. (In fact, the LZ 2 also had such stabilizers, but either they were too small to be of use or the ship never moved fast enough through the air for them to become aerodynamically effective.)

As it turned out, that single change in the design of the LZ 3 would make a substantial difference to the outcome. On its first flight, on October 9, 1906, the ship carried eleven passengers for more than two hours at the respectable speed of twenty-four miles per hour. The craft was fully controllable and did not porpoise up and down or otherwise misbehave in the slightest.

In a letter to the Imperial Chancellor, Zeppelin now bragged that he could build airships "with which, for instance 500 men with full combat equipment can be carried for the greatest distances." The government was so impressed by the new ship's performance, and by Zeppelin's somewhat wild boasts, that it offered him half a million marks for further research and development.

Zeppelin and Dürr made some additional modifications to the craft over the summer of 1907, and by the fall of that year the LZ 3 was making flights up to three hours long and carrying passengers, including Zeppelin's daughter Hella, his only child, and, for once, Hugo Eckener himself. On September 30, the LZ 3 flew for almost eight hours straight.

Once upon a time, Zeppelin had claimed that for airships to be useful to the military they would need an endurance of twenty-four hours. The German government now said that if Zeppelin could build and demonstrate such a ship, it would purchase both it and the LZ 3 for 2,150,000 marks.

Two million marks! Clearly, this was progress.

By this time, Count Zeppelin had become a media sensation. News of the LZ 3's eight-hour flight had spread across Germany, and people who had been in the path of the craft had seen, heard, and been awed by this mammoth "flying ship" cruising majestically overhead, a vision that inspired, according to one historian, "a mix of fascination and fear." The craft had ceased to be a mere flying machine and had now "passed into the realm of the sublime."

Soon Zeppelin's name had acquired a trademark status, and a variety of branded artifacts suddenly appeared on the market, including Zeppelin cigarettes, flowers, and various trinkets, along with leather coats "to be worn in the air." From that point onward, Germans would follow the vagaries of Zeppelin's career much as modern-day Americans trace the ups and downs in the lives of rock stars, sports figures, and other pop idols.

The next ship in line, everyone knew, was the LZ 4. The LZ 4 was to be bigger and better than anything that came before it. It would have the same cylindrical profile—not for aerodynamic reasons, but because a straight-sided tube was the easiest and cheapest shape to design and build—and be almost 450 feet long, 42 feet in diameter, and filled with 530,000 cubic feet of hydrogen, parceled out in the now-standard 17 cells. The engines were more powerful than those of earlier models: two big Daimlers, each of them making 105 horsepower and giving the craft a top speed of 30 miles per hour. The craft had elevators and rudders and stabilizing fins—a full array of control surfaces. This was a real, grown-up, honest-to-God flying machine.

Nevertheless, being a zeppelin, it also had its idiosyncrasies. One of them was that in order for the engines to be refueled, they had to be stopped, during which time they provided no power. The gas tanks in each gondola held only so much fuel, and they had to be replenished by hand, with crew members carrying fresh gas in eight-gallon containers, and apparently it was not considered safe to refill the tanks with the engines running and in a thirty-mile-per-hour slipstream.

Stopping the engines did not turn out to be a problem, however, at least not at first. After three trial flights in June 1908, the LZ 4 set out on July 1 for an extended trip over Switzerland. With Zeppelin himself once again in command, and Ludwig Dürr in the front gondola beside him, the flight lasted twelve hours and covered a distance of 240 miles, setting a new endurance record in the process.

Germany's Zeppelinites went gaga over this, and the mania only intensified a week later, on July 8, which was the Count's seventieth birthday. Seventieth birthdays were very big deals in Germany, and to celebrate Zeppelin's people danced and partied in the streets and squares, some of which were even renamed in his honor. In Germany to this day several cities boast a *Zeppelinstrasse* or a *Zeppelinplatz,* if not both. The King of Württemberg awarded Zeppelin the Gold Medal of Arts and Sciences and told him in a birthday message that "at a time of life when most men retire, you are just beginning to reap the fruits of your great work, you are still at the height of your power." Groups of *Gesangvereine* (singing clubs) stationed themselves outside the Count's home in Girsberg, Switzerland, where his wife and daughter were living, and serenaded the Zeppelins all day long and into the night.

Emboldened by his record-setting twelve-hour flight, the Count now decided that the new ship was ready for its marathon twenty-four-hour challenge trip—the voyage that, if successful, would earn him more than two million marks. This was scheduled for July 14, 1908. The big day arrived, and the craft departed Lake Constance. But only a few minutes into the flight a fan blade fell off one of the engines and dropped into the water below, forcing a strategic retreat to home base.

The next day he tried again. This time the ship never even left the hangar. Instead, it smashed into the side of the floating shed while being towed out, damaging not only some of the craft's structural rings and longitudinals but also a propeller, a rudder, and a few of the internal gas cells.

The great day of reckoning finally arrived on August 4. The ship's cells had been inflated with a fresh supply of hydrogen provided by the Prussian Airship Battalion, and this infusion of new and pure lifting agent would give the ship the maximum possible buoyancy for its historic, long-endurance proving flight. At departure, the LZ 4 had a 550-gallon fuel supply, good for 31 hours of flying time, plus 1,450 pounds of water ballast. Twelve people were on board, including Ludwig Dürr, and of course Count Zeppelin himself.

And so, at a little after six in the morning, with Hugo Eckener standing by on the support float, Zeppelin gave the order "Up ship," a crew member released 132 pounds of ballast, and the LZ 4, it was said, "went up like a curtain."

The plan was to fly along the Rhine River valley to the city of Mainz, and then return home by the same route. The public would follow the ship's progress by telegraph, telephone, and word of mouth and through newspaper accounts.

Initially, to conserve hydrogen, the ship flew at a low altitude of only about 300 feet. It did this by maintaining a nose-down angle of attack, an attitude that produced a given amount of aerodynamic downforce, or negative lift. But when, after seven hours of flight, the forward engine had to be stopped to refill its gas tank, the ship's speed decreased and the nose-down attitude, and resultant downforce, could no longer be sustained. The ship consequently floated up to 2,700 feet, which is where its problems began.

The fact was that flying an airship was not much like flying a conventional fixed-winged aircraft, which was a relatively straightforward business by comparison. An airship was far more complicated. For one thing, in an airship the expansion and contraction of the hydrogen inside the gas cells always had to be taken into account. As the ship rose to a higher altitude, progressively less atmospheric pressure was exerted on its gas cells. But that decrease in pressure would allow the cells to expand, and perhaps even burst, unless some of the hydrogen gas was released. Moreover, on long flights, such as were envisioned for the LZ 4, even the use and depletion of gasoline supplies made the ship lighter, and this effect also had to be counterbalanced by the release of hydrogen.

Further, the hull's constant exposure to sunlight heated and expanded the volume of hydrogen gas, requiring that yet more of it be valved off in order to prevent the cells from bursting. To compensate for the lost hydrogen, it was necessary either to drop some water ballast or to fly at a slight nose-up angle in order to create an amount of aerodynamic lift equal to the amount of lost aerostatic lift. Flying at a nose-up angle, however, produced increased drag, which in turn reduced the craft's airspeed and somewhat retarded its progress. The business of flying an airship was a delicate balancing act with a progressively increasing cascade of ripple effects.

An hour later the whole refueling process had to be repeated for the aft engine. Airspeed bled away again, the craft drifted higher up, and additional hydrogen was released from the cells and wafted off into space.

The beginning of the end was not far off, and it was announced by the failure of a radiator fan gear. The Count decided to land on the Rhine and make repairs. After having done so, however, it was evening and much cooler. The hydrogen aboard was now denser, and dense hydrogen is heavier and has less lifting power. In fact, the craft was now too heavy to rise from the water. At that point Zeppelin decided to treat all nonessential gear as excess ballast, and so the crew heaved overboard whatever they could do without: empty gas cans, extra ropes, as well as themselves, since many of the crew members were nonessential as well, and in effect just so much dead weight. They too were "ballast" and had to go. Even then, a tow from a passing steamer was needed in order to launch the craft back into the air.

At about 10:20 p.m., the LZ 4 finally became unglued and lurched up off the lake. One of the engines then threw a fan blade, followed shortly thereafter by the remaining engine's giving out when its crankshaft bearing melted. The powerless and inert airship was now being blown backward like the free balloon it had suddenly become. The flight had all but turned into a replay of the ill-fated flight of the LZ 2, the craft that ended up looking like (in Eckener's words) "the skeleton of a giant whale."

But in this case, by sheer coincidence, the zeppelin was within range of Stuttgart, site of the Daimler motor works, manufacturer of the craft's failure-prone engines. Why not land there and let their mechanics fix the damn things?

At about eight o'clock the next morning, Zeppelin put the ship down safely in a field near the soon-to-be-famous town of Echterdingen. Crowds gathered around to see the great machine. Soon a contingent of Daimler technicians arrived and started in with the repair work.

And then, as if preordained by fate, the inevitable happened: a storm came up, and with it high winds. An object whose outer dimensions were 450 feet by 42 feet was in effect a kite, or a sail, presenting 18,900 square feet of area to the slightest breeze. A group of soldiers who had been standing by to guard the craft tried to hold on to the leviathan by its mooring lines, but they were no match for the stronger gusts.

One of the tragic flaws of the zeppelin now became clear: there was a basic dimensional mismatch between the craft itself and its makers and keepers. The ship was so immense that it was too big for the size and strength of its handlers and crew. And so the 450-foot-long airship, filled with a huge quantity of Henry Cavendish's "inflammable air," was pushed sideways across an open field. It struck some trees, whose branches pierced the hull and ripped open a gas cell.

The cells were made of rubberized cotton. As Count Zeppelin and the others would belatedly learn, layers of rubber moving against each other are capable of producing sparks of static electricity. There was often some ambient hydrogen wafting around inside an airship, and a spark of static electricity was all it took to set the stuff ablaze. That apparently was what happened next to the LZ 4: flames burst forth from the bow of the craft, and within a matter of seconds the entire ship was engulfed in fire. A minute later the LZ 4 had been reduced to a charred wreck strewn across the landscape.

It is controversial whether any deaths or injuries were involved. An account purportedly from "an eye-witness of this tragedy" claimed that "one of the mechanics was fatally injured by his fall." A later report said: "One of the mechanics and two onlookers, wounded by the explosion, were taken to Stuttgart, but no one

Zeppelin LZ 4, August 5, 1908, Echterdingen, Germany.

died," while a third stated that three men "luckily escaped with minor injuries."

Hugo Eckener once again traveled to a zeppelin crash site. "The burnt skeleton of the airship lay dark and massive in the field like some slain monster," he said. The man had a gift for words, no question.

In any event, this was *Hindenburg* prequel number 3.

By any reasonable standard, the total destruction of this latest aerial monstrosity might well have sounded the death knell for the zeppelin, as well as prompting a polite good-bye to the man who had created it. But no, just the reverse happened. Instead of being dismissed as a nut, public nuisance, menace, crank, or crackpot, Count Zeppelin was now embraced as a hero, a star of the age. And the airship, instead of being seen as a hydrogen bomb waiting to go off, assumed the status of an almost divine object, one that, as a matter of German honor, would have be made into a good and proper icon for eternity.

Thus began the Delirium.

The Delirium was a mass hallucination, shared psychosis, or induced delusional disorder. This strange phenomenon amounted to a nationwide *folie à deux,* one that affected all age groups, both sexes, and all classes of society: rich and poor, rustics and urbanites, royals and commoners. It was as if a universal mass drugging had occurred.

The symptoms of the malady were, first, the belief that the LZ 4 pileup and auto-da-fé was not a disaster but rather a miracle—specifically, "the Miracle at Echterdingen," a ringing apotheosis of the *Kolossal*. Count Zeppelin himself, furthermore, was no longer "a genie flying a liverwurst," as a journalist had once dismissively referred to him. He was not even any longer a man, a mortal, or even just a count: he was now a symbol, a folk hero, an icon, an idol, virtually a saint. His invention, faulty as it might have appeared to a few philistines in its first few simple, naive incarnations, was in reality an emblem of national pride and aspiration, something that would work—that must be *made* to work—against all odds, despite everything, and at all costs, no matter what. The incident at Echterdingen was therefore not an ending but a beginning—the beginning of Teutonic supremacy in the skies.

And so, overnight, there arose a vast outpouring of grassroots sentiment and support for Count Zeppelin and his project. Children broke open their piggy banks and sent their precious few pfennigs to the Count. There were contributions from the well-to-do, "from savings under beds, from Germans living in faraway lands, from the bowling club at Baden who sent 150 marks and the mining association of Essen who sent 100,000," historian Douglas Botting reported. "Those who had no money sent sausages, hams, wine, a pair of hand-knitted woolen socks." At any public gathering where the zeppelin or Zeppelin happened to be mentioned, people spontaneously burst into passionate renditions of the *Deutschlandied* (aka *Deutschland über alles*), the stirring hymn by Franz Joseph Haydn that was later to become the German national anthem.

Soon came the souvenir items and the semi-mythic totem objects: Zeppelin-decorated cigarette cases, dolls, harmonicas, yachting caps, airship-shaped lamps, and, absolutely the most fitting of them all, a figurine of the Count in the form of a nutcracker.

On November 10, three months after the disaster, the Kaiser himself, Emperor of Germany, Wilhelm II, came to Friedrichshafen, home of Zeppelin and the zeppelins. Royally outfitted in an immense green cape, feathered hunting hat, and walking stick, the Kaiser awarded Count Zeppelin the Prussian Order of the Black Eagle, the highest mark of chivalry in Prussia, and lauded him as "the greatest German of the twentieth century"— the century, at that point, being all of eight years old.

Within a few days of the immortal Echterdingen event, the Count was fairly rolling in dough. He had received contributions totaling more than six million marks—all thanks to another demolished and incinerated example of his accident-prone aerial disaster-machine. Zeppelin, who yet once more had been ready to vow that he'd *never again build another airship!* had now been stricken anew by a fresh case of zeppelin fever.

At this precise stage in his career, Count Zeppelin's airships— despite their extreme size and the fact that some of the funding had come from state lotteries and from outright government grants—had amounted largely to a do-it-yourself, homebuilt project that was in fact a personal crusade. Zeppelin had conducted his operations more on the model of a hobby or avocation than as a formal business enterprise. That would change in the wake of the huge financial windfall resulting from "the Miracle at Echterdingen." There was suddenly just too much money in his hands for him to put to effective use on his own.

The first item on his agenda, then, was to convert the process of airship design and construction into a legitimate and incorporated business enterprise. And so, on September 3, 1908, Zeppelin established the Luftschiffbau Zeppelin GmbH (the Zeppelin Construction Company, Ltd.), with headquarters at Friedrichshafen. He allotted some three million of his miracle marks to this venture. Separately, on December 30, the Count created another legal entity, the Zeppelin Foundation (Zeppelin Stiftung), for the purpose of overseeing distribution of the rest of his instant manna from heaven, essentially another three million marks. And it was now clear also that the Count would need outside professional help in running these organizations—in other words, a business manager. To occupy that position he chose Alfred Colsman, son-in-law of the aluminum producer Carl Berg.

Colsman was tall and thin, and he dressed, thought, and acted like a banker. He had studied business and economics at Berlin and Geneva, and he brought to the Zeppelin Construction Company the perspective of a hard-nosed entrepreneur. One of his first official acts was to replace the old floating construction shed on Lake Constance with an updated, and larger, shore-based version. And so there arose on a tract of land north of the city an even more impressive construction shed, a double-wide iron hangar that was capable of holding two colossal airships at once, side by side. It was a building of enormous dimensions: 584 feet long, 151 feet wide, and 65 feet high, a gargantuan beast in its own right. It would in time become a zeppelin factory, churning out the next new wave of airships, craft that, Zeppelin and Colsman hoped, the military would be interested in buying.

Colsman would also need a staff of his own, including someone who could advertise and tout the virtues of this new line of oversize flying machines. And who was better suited for the task than Hugo Eckener? Ever since his first meeting with the Count in 1900, Eckener had been mystifyingly starstruck by zeppelins. Then again, he was only human. And so, at Colsman's invitation,

Eckener now became the public relations man for the Zeppelin Construction Company.

Above and beyond these strictly business matters, there were more serious problems concerning the zeppelin itself. The product as it stood left much to be desired. The ships to date had rained down parts as they flew. Their engines were inadequate, underpowered, and unreliable. The tasks of maneuvering the craft safely while on the ground, especially during takeoff and landing procedures, had yet to be mastered. And then there was the ever-present risk posed by the craft's rubberized gas cells, which, as had been amply demonstrated at Echterdingen, could set off electrical sparks by the simple act of two adjacent rubber layers moving against each other. So the company started experimenting with an alternative material known as goldbeater's skin.

In an aircraft that was already an extremely odd duck on any number of counts, the prospect of making its gas cells out of goldbeater's skin had a very high strangeness quotient all by itself. Goldbeater's skin was a specialized and comparatively rare product: it was the outer membrane of a portion of cattle intestine called the cecum, which is a pouch attached to the junction of the animal's large and small intestines. The membrane had been used, perhaps since antiquity, to make gold leaf, which goldbeaters produced by starting out with a small ingot of gold and beating it thinner, then sandwiching the leaves between sheets of cow intestines, ultimately creating a stack of alternating goldbeater's skins and gold leaf that could be more than 100 layers deep.

Goldbeater's skin was used for that purpose because it was light, thin, elastic, and exceptionally tear-resistant. It was also virtually impermeable to air, and even to hydrogen, whose molecules were the smallest of all the elements. Count Zeppelin had probably heard about goldbeater's skin from tales that went back to the first days of ballooning, to the time of the Montgolfier brothers and J.A.C. Charles in the summer of 1783, when Parisian balloonists were making use of the membrane.

But goldbeater's skin came at a substantial price. A cow's cecum, after all, was a small thing (at least as compared to a zeppelin): it consisted of a cylindrical sac that when split open and laid flat was only about a foot wide by two feet long, about as big as a dishtowel, and as many as 15 skins were needed to prepare one square meter of the stuff. A single gas cell could require up to 50,000 skins, and there were 17 gas cells in a zeppelin, meaning that up to 850,000 skins (and as many cows) could be required to make a complete set of gas cells for a single airship. As aviation writer Ernest Gann once said, 850,000 cows was, "even by Texas standards a lot of cattle off the range."

While the Zeppelin Construction Company's engineers were experimenting with goldbeater's skin, they proceeded to build a succession of new craft that, each in its own special way, flew headlong to an all but inevitable and predetermined doom. The LZ 5, for example, was practically a carbon copy of the luckless and self-immolated LZ 4. But Count Zeppelin wanted to show off the new craft's capabilities by making an epic thirty-six-hour demonstration flight to Berlin. On the first try, it ran into a rainstorm that made the ship so heavy that the crew was forced to drop all available ballast and return to base. The second try ended, after thirty-seven hours, when Ludwig Dürr, who was at the controls, fell asleep on the approach to landing and flew the ship into a pear tree. The impact destroyed the nose and deflated two forward cells, but for unknown reasons caused no fire.

The army nevertheless took delivery of the ship. But on a later flight to Cologne the new military crew was compelled to make

"Gone with the wind": Zeppelin LZ 5, April 24, 1910, Weilburg, Germany.

an emergency landing due to engine trouble. While aground, winds tore the airship from the hands of the soldiers responsible for holding it down; the craft skimmed across the landscape and ended up as a lurid wreck on a nearby hillside.

One might imagine that this almost unbelievable run of bad luck—and all of it, moreover, on the physically grandest possible scale—would have to end with one craft or another, sooner or later, eventually. But the very next ship in sequence, the LZ 6, was not that craft. On September 14, 1910, it burned up while in the comparative safety of its hangar when one of the mechanics was cleaning the gondolas with gasoline while another one started an engine, which caught fire, and a third poured a can of gasoline over it, probably thinking that it was water. It was a performance fully worthy of The Three Stooges, the mid-twentieth-century slapstick comedy team. Anyway, that was the fate of the LZ 6, a ship that turned out to be "worth more to her owners dead than alive," said historian Douglas Robinson. "Insurance payments of 320,000 marks from Lloyd's of London gave the Zeppelin enterprise a much needed 'Shot in the arm.'"

Still, since it too had ended up in flames, the LZ 6 was also *Hindenburg* prequel number 4.

With this latest disaster, four out of the six airships produced by Zeppelin and company had terminated their lives as wrecks. The very first, the LZ 1, had escaped violent destruction. After its third and final flight, it had been deliberately and politely dismantled, piece by piece. A similar fate awaited the LZ 3.

Now, with neither the German army, navy, nor anybody else in the market for airships, Alfred Colsman had the idea of starting a zeppelin *airline* as a way of cashing in on the Delirium, which was still going strong with the general public. Instead of carrying bombs, troops, ammunition, and equipment, zeppelins would fly paying passengers.

Although the Count himself didn't much like this scheme—he considered it a perversion of the military role that he had envisioned for his airships—a number of German city fathers did. As did Albert Ballin, director of the Hamburg-Amerika Line, which would acquire exclusive rights to sell zeppelin tickets. And so, on November 16, 1909, Colsman and Ballin cofounded the airline that came to be known as DELAG, the acronym for its lengthy German name: Deutsche Luftschiffahrts Aktien Gesellschaft (the German Airship Transportation Company).

It was the world's very first airline—and it was a courageous (or reckless) one at that, given the fates of the four previous examples of ships that were now to carry paying passengers, who would in effect be staking their lives on a possible, even probable, next succession of incipient wrecks. The world's very first airline would also boast the world's very first commercial airliner, the LZ 7, which Colsman, as head officer of DELAG, now ordered from himself, as managing director of the Zeppelin Construction Company. Perhaps this craft, with "lucky 7" going for it, not to mention the law of averages, would succeed where the others had failed.

Grandly named the *Deutschland* (a name that Zeppelin had first applied to his notional and finally abandoned "aerial express train"), the LZ 7 would be the largest zeppelin built to date: 486 feet long, 46 wide, and with a fanlike multiplicity of rudders at the stern, plus stacks of elevators fore and aft. If anything, the ship was oversupplied with control surfaces. The craft had space and lift enough to carry twenty-four passengers amidships, in a mahogany-paneled, carpeted cabin that was further enhanced by mother-of-pearl inlays on columns and beams. It sounded like a flying whorehouse. When it crashed—there seemed to be no "if" about it—at least it would make for an elegant carcass.

To deflect attention from the appalling track record of the ma-jority of zeppelins built to date, DELAG advertised the new ship as a masterpiece of solid construction, with the cabin framing "braced so securely in position by twelve pairs of steel cables that it will hang immovable from them even if, through some unlikely accident, a couple of struts should bend or break."

The historic first commercial flight of the *Deutschland*—which would also be the first commercial flight of the world's first airliner—was scheduled to take place on June 28, 1910. For this momentous occasion, the passenger list would consist exclu-sively of twenty-three journalists, who would be wined and dined aloft by the twenty-fourth passenger, DELAG managing direc-tor Alfred Colsman. Colsman must have considered this a public relations coup, the kind of advertising that you just cannot buy for any money.

The flight unfolded in a manner that, for a zeppelin, was par for the course. Which is to say that shortly after takeoff the ship encountered winds that were stronger than what its engines could overcome, so that the craft found itself being blown backward over the usually lovely but now suddenly forbidding Teutoburger Forest of northern Germany. Updrafts from a thunderstorm in the vicinity lofted the *Deutschland* to the heady altitude of 3,500 feet, with the gas cells continuously and automatically venting hydrogen during the ascent. The consequent loss of lift, added to the significant weight of the rain that now started to pour down heavily upon it, forced the ship toward the ground.

The craft had no radio, nor any other means of communication except for yelling, and so as a last resort the captain, a conscript from the Prussian Airship Battalion in the Count's absence (Zeppelin was on an exploratory trip to Spitzbergen, scouting out possible flight routes to the North Pole), took to dropping handwritten messages over the side, requesting any and all able-bodied men nearby to help land the ship. In the event, no such help was needed because instead of landing, the zeppelin slammed into the forest treetops, an arboreal perch from which the stunned journalists now had to climb down some thirty feet to the forest floor below. In the next few days, a wrecking crew arrived with the usual saws and axes. So much for free advertising.

But DELAG, like Count Zeppelin himself, harbored an aptitude for bouncing back. The Zeppelin Construction Company now cloned into existence a replacement ship, which they christened the LZ 8 *Deutschland II*. The craft was to be based at Düsseldorf, which was a strange location in the circumstances because the surface winds there were known to be gusty and altogether unpredictable. The airline now made safety job one, however, and so, in an effort to protect their new craft from possible crosswinds while exiting or entering the hangar, they erected a 30-foot-high wind-deflecting barrier that extended from the hangar doors outward for 150 feet onto the airfield. Still, 150 feet was not even half the length of the 486-foot-long *Deutschland II*.

One of the first commercial flights of the *Deutschland II* was set for May 16, 1911, under the command of Hugo Eckener, despite the fact that Eckener had never before piloted an airship and had no particular credentials for doing so. As before, the company tried to reassure potential passengers that zeppelin travel was really not as dangerous as they might have imagined. Their new advertising brochure noted that "most people will believe that it takes great courage to entrust themselves to the gondola of an airship," but went on to allay such fears. "The ascent of an airship takes place in such unexpected and complete tranquility that the passenger literally will not be aware of the motion if he shuts his eyes," and so on.

This rhetoric must have worked its magic, because on May 16 a full load of passengers boarded the craft while it was safely enclosed in its Düsseldorf hangar. A massive ground crew (estimated at 300 men) slowly walked the ship from its shed. And then, in a twist of fate so absolutely characteristic of zeppelins as to be paradigmatic, no sooner had the aft end emerged alongside the protective windscreen than a crosswind lifted the ship up and over it, taking some of the ground crew with it, and planted the stern on the top of the wind deflector. The middle section, meanwhile, came to a rest atop the front end of the hangar roof. The craft hung there suspended and pointed upward, like a rocket to Mars.

LZ 8, *Deutschland II* at Düsseldorf, Hugo Eckener commanding, May 16, 1911.

The passengers climbed down a fire ladder that reached from the passenger cabin to the ground—a giddy descent, and not for anyone with a fear of heights. Such was the short life and sudden death of the LZ 8 *Deutschland II*.

At least there would be no *Deutschland III*.

Chapter 5

DEMYSTIFYING
GARGANTUA

B y the time the LZ 8 *Deutschland II* had planted itself so pic-
turesquely atop the Düsseldorf hangar in 1911, Count Ferdinand von Zeppelin had been building zeppelins for more than
a decade. During that period, the process of designing and constructing an airship had become somewhat routinized, even insti-
tutionalized. The truly *Kolossal* size of these monsters, especially
the super-king-size *Graf Zeppelin* and the *Hindenburg,* could foster the impression that these ships were miracles of creation, or
that some very special magic tricks or heroic engineering feats
were required in order to put them together.

But that impression would be misleading, as there was far less
to a zeppelin than met the eye. To begin with, for all of its grand,
extra-large, jumbo, and otherworldly immensity, the inside of a
zeppelin was mostly empty space. And paradoxically, it was actually emptier when its gas bags were fully inflated with hydrogen
than when they were deflated and the empty space was taken up

by ordinary air, the reason being that hydrogen was less dense than air. That lesser density, of course, was precisely what gave the hydrogen-filled zeppelin its lift. So if you think of an air-filled zeppelin as "empty," then a hydrogen-inflated zeppelin was even emptier—*emptier* than empty. The interior of the *Hindenburg* was in that sense mostly nothing, a lacuna approaching the limiting case of the true and genuine nothing of Francésco Lana de Terzi's spheres of vacuum.

That explained why the hydrogen-filled *Hindenburg* could be destroyed by fire in the space of thirty-four seconds: despite its stupendous overall dimensions, there was comparatively little inside the craft other than the seven million cubic feet of exceptionally thin, insubstantial, and enormously explosive hydrogen gas (and the seventeen cells that contained it). Once that flammable substance had been ignited, the whole of it was consumed almost instantaneously. By way of contrast, had the inner recesses of the *Hindenburg* been as materially dense as the inside of a battleship, or even an office or apartment building, it could never have vanished from the face of the earth in just half a minute.

Furthermore, although an airship's rigid framework enclosed a large, yawning, empty chamber, the framework itself was composed of parts that were individually relatively small and easy to design, produce, and assemble into larger structures. Indeed, there was an obvious commonality between the way zeppelins were built and the growth of the largest terrestrial organisms, such as elephants, whales, and dinosaurs, which, like the zeppelin, came into existence part by part through the successive accretion of individually insignificant building blocks over the course of time.

A zeppelin's basic unit of construction was the aluminum (or aluminum alloy) beam or girder. In the first zeppelin, the LZ 1, the girder, whether it was part of a horizontal, longitudinal member or a transverse ring, was made up of an I-beam that itself was composed of top and bottom T-sections facing each other about seven inches apart, held together and cross-braced by diagonal struts in the form of an X. Individual sections of these unit structures could be joined together lengthwise, one to the next, to form a longitudinal girder, or they could be attached to each other at shallow angles to make up a transverse ring.

A transverse ring was a twenty-four-sided regular polygon that from a distance approximated a circle. There were main rings and secondary rings. The main rings were braced by wires that ran diagonally from one side of the ring to the other, like the radial spokes of a Ferris wheel, as well as by separate chordal wires that ran in straight lines between two relatively close portions of the ring arc. The secondary (or intermediate) rings were not wire-braced. The gaps between two main rings formed an open space or compartment in which a single gas cell could be hung, confined, and inflated.

As a final complication, all of the rings, whether main or intermediate, were joined to the longitudinal girders that ran horizontally for almost the entire length of the ship (except at the nose and tail sections, where they were inwardly tapered to endpoints). One of the more complex structural elements of the airship's framework was the intersection of the longitudinal girders and the ring girders. This was essentially the engineering problem of connecting a straight I-beam to a ring that ran at right angles to it. Conceptually, this was simple enough—it was like attaching a hoop at right angles to a stick—but the physical implementation of the spot where the two met was a joint. Ludwig Dürr produced a variety of different geometries for that pivotal junction over the several years of airship design and construction.

The basic aluminum unit sections were fabricated in the Carl Berg factory at Lüdenschied, in northern Germany, and were then

crated and shipped to the zeppelin construction shed on the shore
of Lake Constance. Each individual ring was first pieced together
flat on the floor of the construction shed; when complete, it was
then temporarily attached to a separate "assembly ring" (also called
a *Montagering,* or mounting ring) that raised the newly built transverse ring from a horizontal to a vertical position. According to
Dürr's own account of the assembly operation, "The rings are then
moved along rails in the roof of the shed into their final position
and hung up from their suspension eyes. The building process
is speeded up by assembling and hanging a central section, then
working towards the ends."

Early zeppelins were cigar-shaped not because this was an aerodynamically optimal or even efficient form—it wasn't—but because the uniformity of cross-sections allowed the ship to be built
out of simple reiterations of exactly the same ring structure. This
both sped construction and reduced design and fabrication costs.

As the airship grew a wooden scaffolding and external support
system grew up around it, and this formation got progressively
ever larger, until the combination of airship parts and supporting
elements looked like an intricate mazeworks of aluminum, wood,
wires, and fabric. Adding to the clutter were various cables, ropes,
pulleys, stairways, walkways, cantilevers, towers, and stationary
ladders, plus a collection of tall, wheeled fire ladders that workers
who evidently had no fear of heights climbed to gain access to the
various external sections of the ship.

The construction crew members assembled the nose and tail
structures as separate modules and then attached them to the
finished cylindrical hull as complete units. With the rigid framework now complete, they added on the further components of
the airship, chief among which were the usual sixteen or seventeen (or in some cases as many as nineteen) gas cells. A wire-
netting meshwork enclosed each cell to form an external limit for

expansion and to distribute the lifting loads from the top of the cells downward and evenly to the rest of the framework.

Among the essential parts of a gas cell were nozzles for filling and deflation, as well as valves, both automatic and manually operated, for the release of hydrogen gas, either to relieve excess pressure or for maneuvering purposes, i.e., to permit descent. Ductwork to vent the released gas was also installed on some airships, whereas on others the hydrogen could escape through the porous outer covering. In its passage through the fabric, the gas "was well diluted with large amounts of air leaving no inflammable mixture outside the airship," Ludwig Dürr explained. "This was an important factor in complying with a specification of the military authorities that the machine guns on top of the airship should be free to fire at all times."

In the case of airships not equipped with machine guns or other armaments, the craft's subsidiary structures and systems— engines, propellers, power transmission shafts and gearing, fuel tanks and fuel lines, gondolas, passenger accommodations, steering apparatus and cables, instrumentation, and whatever else— were added on progressively to the hull framework. All of these subsidiary structures were arranged around the perimeter of the craft, so as to leave the interior largely "empty."

The ship's outer covering gave it a relatively smooth appearance, and helped produce a laminar airflow, while also protecting the gas cells and other internal components from the elements. "In the early days the cover was attached to the hull in vertical strips one bay wide," said Dürr. "These strips were hooked to the ridge girder and rings with clasps, and laced together along the lowest longitudinal girder."

All of those vertical strips created a lot of surface drag, however, and so later zeppelins had their covering applied in *horizontal* strips, with the seams between them pasted over with narrow pieces of glued fabric. In very late model airships, such as the *Hindenburg*, the outer surface was painted with a mixture that included powdered aluminum to help deflect the sun's rays and to reduce the heating of the gas cells. It also made the airship into a fairly luminous object in sunlight, thereby enhancing even further its already ethereal appearance.

The "Miracle at Echterdingen"—that is, the LZ 4's destruction in August 1908—had been followed by an unbroken string of zeppelin accidents, disasters, or other calamities: the LZ 5 wrecked and sprawled across a Weilburg hillside; the LZ 6 burned to a charred remnant inside its hangar; the LZ 7 *Deutschland* marooned on the treetops of the Teutoburger Forest; and the LZ 8 *Deutschland II* precariously balanced between its hangar and its protective wind-breaker at Düsseldorf. With five airships in a row dead and gone, three of them in a single year, 1910, it was a triumph of optimism over good judgment that nobody in the airship business saw the writing on the wall and proposed abandoning the whole fairly insane lighter-than-air enterprise.

But no one did. In the summer of 1910, in the immediate aftermath of the LZ 7's Teutoburger Forest incident, Alfred Colsman, with no apparent awareness of incongruity, appointed Hugo Eckener as the airline's flight director, making him responsible for recruiting zeppelin pilots and teaching them the basics of flight, particularly flight safety. Inasmuch as Eckener had had no prior experience in flying airships, although he had been allowed to touch and even operate their control wheels on rare occasions, his selection for the post was somewhat odd, as Eckener himself was the first to admit. The appointment, he confessed, "caused me many sleepless nights and tortured hours." Why he ever accepted the position in the first place was something of a mystery.

Equally mysterious was the commitment by the Zeppelin company and DELAG's directors, even in the face of the five successive destroyed or dismantled airships, to continue building and flying the craft. As Eckener later explained it, "The directors, many of whom had a strong technical interest in the matter, had now been, so to speak, 'bitten' by the problem and felt the more impelled to continue the battle along Count Zeppelin's lines since great progress had just at that time been made in the matter of engines, and much had been learned which might serve to avoid further disasters."

"Bitten" by the problem—meaning, in essence, that the emotional and cognitive spell cast by the zeppelin continued to exert its magical effect on the German airship fraternity. But just at that crucial point, it so happened that the otherwise normal and invariable string of zeppelin debacles was broken by a period of idyllic stability and calm. The first of these beatific eras commenced with the maiden flight of the LZ 10, the *Schwaben* (named for Swabia, the region of Germany where the Count had been born and the zeppelins were built). The craft has been called DELAG's "lucky ship," and with good reason, for it did not immediately implode, explode, burn up, fall to pieces, or meet with any other misfortune. In fact, the ship made hundreds of flights without a single mishap. But luck was by no means the whole explanation. To begin with, the LZ 10 was slightly smaller and therefore marginally easier to handle near the ground than its immediate predecessor, the LZ 8. (The numerically next LZ 9, a military airship, came off the assembly line out of order, after the LZ 10.)

Second, many of the passenger flights of the *Schwaben* were conducted in the summer of 1911, which was a period of uncharacteristically placid weather, in conditions that were tailor-made for zeppelins: blue skies, calm winds, and a relative absence of destructive storms. Eckener himself later spoke nostalgically of "that marvelous summer of 1911 with its three months of practically unbroken fine weather, a perfect godsend," which he in fact attributed to a "higher power."

Also explaining the luck of the *Schwaben* was the fact that the craft was propelled not by the underpowered and sickly Daimler engines but by three 145-horsepower—and, for a change, comparatively reliable—Maybach engines, power plants that had been designed and manufactured specifically for airships. Further, the *Schwaben* was to be stationed at the airship base at Baden-Oos, which was located in a sheltered valley and out of the winds. And last of all, there had even been an innovation in the realm of ground handling: Eckener, who was always on the lookout for better ways of doing things, had proposed a system of docking rails that would extend from the hangar to a distance of 600 feet out onto the airfield. The airship would be attached by cables to trolley cars that ran along the rails, holding the ship securely to the ground until it was ready to be launched. The apparatus must have worked, for the *Schwaben* is credited with making 364 flights over the space of a year, carrying more than 6,000 passengers without incident.

Nevertheless, even the *Schwaben*'s fabled good luck finally ran out, on June 28, 1912, when, like many of its predecessors, it was torn from its moorings by a gale at Düsseldorf. According to one account, "*Schwaben* caught fire above the forward gondola almost immediately and was quickly entirely engulfed in flame."

It is unclear whether anyone was injured in the accident: no passengers were aboard at the time, but contemporary news reports claimed that from thirty to forty people (presumably ground crew members) were injured, along with its designer, Ludwig Dürr, who had been piloting the craft. Later accounts claim there were no injuries. Whatever the truth, an investigation determined that the fire was caused by sparks produced from friction between its rubberized gas bags.

And so the LZ 10 *Schwaben*, "lucky" as it had been for a time, in the end succumbed to its preordained and inevitable fate, thereby becoming *Hindenburg* prequel number 5.

It was at this point that the Zeppelin Construction Company got serious about fabricating its gas cells out of goldbeater's skins. This was no trivial matter, however, for the preparation of even a single piece of goldbeater's skin was an arcane, time-consuming, and labor-intensive process, and thousands of such skins would be required to make even so much as a single gas cell. The skins, after all, were the thin outer membranes of cattle intestines. These body parts would be supplied en masse by butchers and slaughterhouses, but that was only the beginning. The membrane first had to be separated by hand from the rest of the intestinal wall. Then it had to be washed and cleaned of blood, mucus, fat, grease, and other offal. This required a certain amount of messy scraping with a blunt knife or other instrument. Then the membrane had to be soaked in a salt solution and rinsed in running water. As one practitioner of the art described it, "The processing of cattle caecum is necessarily an unsavory business."

Workers stretched the cleaned skins on boards and dried them. Double-layered skins were made by placing one skin over another one of the same size before drying; this made for an exceptionally gastight piece of material. In either case, whether single or double, the end-product was a section of skin that measured roughly one by two feet.

Then you had to attach one skin to the next, which was done by overlapping them slightly at the edges and gluing them together, the object being to make a seamless and hydrogen-proof closure. Sooner or later you had an actual *sheet* of goldbeater's skin in front of you, and these sheets had to be made into still larger panels by stitching and gluing, and the whole process repeated again and again until there eventually arose a stretch of fabric the size of a zeppelin gas bag.

The assembly process was not perfect, and some skins had holes that had to be patched, bubbles that had to be pressed out, or other defects. Finally, when the goldbeater's skin gas cell was completed, it had to be hung up and tested for gastightness. But even when it was inside the airship and filled with hydrogen, the cell was still vulnerable, as the skins could be, and were, eaten by mice. And although goldbeater's skin gas cells did not produce static electrical sparks when rubbed together, there were other ways of generating sparks, as the Zeppelin company would learn in due course.

DELAG, meanwhile, continued its "airline" operations with a number of new ships. At this point the company did not function so much in the manner of a latter-day airline, in the sense of offering a regularly scheduled passenger service between various cities—the airships were still too susceptible to wind and weather for that—but rather as a sightseeing line, taking passengers on local joyrides that lasted for an hour or two before returning them to the airship's point of origin. Exactly where any given flight went on any given day (or time of day) was more a function of the prevailing wind direction than of any other factor.

DELAG was an unusual "airline" in another sense as well: most of its passengers flew for free. Of the 1,500-plus flights made before World War I, DELAG carried more than 34,000 passengers. But of these, only 10,107 had paid actual folding money for the privilege; the rest of the travelers—the media stars of their day—had been carried for publicity reasons. In 1912 DELAG hired its first flight attendant, Heinrich Kubis, who inaugurated the practice of serving drinks and snacks to passengers.

All that changed on August 7, 1914, at the onset of World War I, when the German army took over the ships for use in battle.

I t was in wartime, supposedly, that the airship would come into its own, its standout virtues shining forth in clear relief. As it turned out, just the reverse happened. For although it had been Count Zeppelin's idea practically from the beginning that airships would be revolutionary instruments for waging war, that notion was roundly overturned by the course of actual events. Zeppelins were even worse disaster-prone, lethal failures in battle than they had ever been beforehand.

The war would prove that zeppelins were generally ineffective as bombers. They were so large and lumbering that they made easy targets for fighter planes and ground-based antiaircraft artillery fire—they were virtual magnets for enemy bullets. In a vain attempt to make themselves invisible, zeppelins flew at night and during the dark of the moon, or above cloud banks, which made precise navigation and visual targeting difficult or impossible. Indeed, so many misdirected bombs landed in the farmlands of the United Kingdom that the British government wondered whether the Germans were trying to destroy crops and livestock instead of buildings and people.

But there was really no effective way of hiding a zeppelin. They were just too enormous. The very size that made them so attractive to Count Zeppelin, and to later airship designers, builders, and the general public, made them easy targets of opportunity for practically anyone with a weapon.

The fact was that even before the war broke out the first airships taken over by the German army and navy for later use in

wartime continued the ritual practice of crashing and/or burning—
only this time often killing some or all of their military crews in
the process. The LZ 15, renamed Z 1 by the German army (whose
airships bore the "Z" designation), was launched on January 16,
1913, and was destroyed after a forced landing a mere two months
later. The LZ 14, renamed L 1 by the German navy (whose airships
bore the "L" designation), broke apart in a storm and sank near
the island of Helgoland in the North Sea, on September 9, 1913,
killing fourteen of its crew of twenty in what came to be known as
the "Helgoland air disaster." (The *New York Times* commented,
with respect to this incident, that "probably no inventor has met
with more discouragement and setbacks than has the persistent
Count Zeppelin.") On that same day, September 9, the navy's new
L 2 was launched from Friedrichshafen; a little over a month later,
on October 17, an engine fire ignited its hydrogen and the craft
fell in flames at Johannisthal, near Berlin, killing its entire crew
of twenty-eight, in an event that in turn came to be known as the
"Johannisthal air disaster."

The Johannisthal air disaster was *Hindenburg* prequel number 6.

Lethal as both of them were, those two disasters occurred *in
peacetime,* before the war ever started. Once the war began, matters took an immediate turn for the worse as airships got shot down
or were otherwise destroyed by the dozen. The German military
operated 117 airships across the course of the war, including those
made by the Zeppelin Construction Company as well as others
made by competitors such as the Schütte-Lanz Airship Company
(Luftschiffbau Schütte-Lanz), of Mannheim. Of that total, thirty-
nine were gunned down by ground fire or by fighter-plane attacks,
while forty-two more were lost owing to other factors such as
weather, structural failure, or crew error, for a grand total of eighty-
one demolished airships, representing a destruction rate of almost

70 percent of those flown. Most of the crew members—more than 400—had been incinerated inside of their falling, flaming wrecks, making for a death rate scarcely equaled by any other branch of the military forces of any country that took part in the war.

From the start, the entire zeppelin military effort had developed into a cat-and-mouse game in which one side devised a strategy, device, or ruse that was soon met by an enemy countermeasure in an ever-escalating war of wits. For example, in an effort to evade ground-based searchlights and nighttime antiaircraft fire, the Germans painted the undersides of their zeppelins black. The British responded by using stronger searchlights and by shining as many as twenty or more of them skyward from different angles so that, according to one zeppelin commander, "the night sky was illuminated as brightly as daylight." Even *without* searchlights, the airships were so huge that they were generally visible anyway because their immense silhouettes blotted out a large portion of the background overhead, whether stars or clouds.

To get around this the Germans flew through fog or *over* cloud banks that even searchlights couldn't penetrate. The problem with that strategy, however, was that the crews couldn't see the ground and therefore often couldn't tell even approximately where they were, whether they were over land or sea, city or country—or indeed, over *what* country. In one instance, the commander of an airship who thought he was dropping bombs over London was actually flying over Arras, France. In another case, a zeppelin looking for Edinburgh in dense fog mistakenly bombed a castle in the Scottish highlands, ran out of fuel, drifted north toward the Orkneys, and finally crash-landed in the waters of a Norway fiord.

To correct such navigational blunders, zeppelin captain Ernst Lehmann (and others, independently) came up with the plan of lowering an observation car into and beneath the cloud banks and having an observer give directions to helmsmen in the control gondola by telephone. This was a great idea other than for the facts that (a) in trials, the so-called cloud car was so unstable that its occupant was nearly tossed out, and (b) the steel cable that towed the car also made for an excellent lightning rod.

In the end, the Germans never did solve the airships' wartime navigational problems. Nor their propensity for getting shot down. They tried to escape antiaircraft fire and fighter-plane attacks by building zeppelins that flew at ever higher altitudes, often exceeding heights of 20,000 feet. That tactic had a number of unwelcome consequences, however: water ballast turned to ice, compass needles froze in their damping liquids, fuel oil congealed in its tanks and lines, engines stopped, crankshafts broke, propellers flew off, and some airships, as free balloons, sailed off to nowhere and were never seen or heard from again.

The great heights at which they flew also wreaked havoc on crew members, who suffered from severe oxygen deprivation and everything that went with it. "Blood ran from the ears, mouths, and noses of the airshipmen," reported Captain Ernst Lehmann, "and vertigo attacked them one after the other. . . . The nervous tension and the extreme cold had rendered some of them almost helpless, and when the mechanics in the engine cars had exhausted their supply of oxygen, they simply lay there half unconscious while the engines ran untended."

The British, meanwhile, developed attack planes armed with incendiary bullets, designed to cause fires on contact, plus incendiary rockets that could fly almost as high as the zeppelins did, thereby denying them nearly every advantage gained by height. In fact, the zeppelins continued to go down in flames regularly, each of them "a roaring furnace from end to end," in the words of British pilot C. S. Iron.

In wartime, therefore, not only was the zeppelin an occasional disaster, it was inherently and systemically an operational disaster, and an expensive one at that: a single zeppelin cost as much as thirty Albatros biplanes. Lehmann tried to put the best face on things with his observation that "even if every third Zeppelin were to be shot down, the crews could still drop their bombs while falling." This, of course, was no consolation to the crews of the machines in question.

Count Zeppelin himself did not live to see the end of the war, the defeat of Germany, and its humiliation by the harsh terms of the Treaty of Versailles. But he had seen his dream turn into a nightmare by the time of his death on March 8, 1917. Far from saving the Fatherland, his airships had turned out to be highly efficient killers of those who flew them into battle.

The Count was buried at Pragfriedhof Cemetery in Stuttgart, under a headstone that bore the epitaph YOUR FAITH HAS HELPED YOU, which was a debatable proposition at best. He was a man who in the end did not truly succeed and did not truly fail: a fanatical believer and wishful thinker, equal parts crackpot, visionary, and buffoon.

For Hugo Eckener, World War I was a learning experience, one that provided him with a vast amount of data concerning the proper operation of zeppelins. He had spent the war as Director of Airship Training for the Imperial German Navy, and had been adviser to Peter Strasser, the leader of Naval Airships. In 1919, on the basis of the information he had distilled from four years' worth of airship combat operations, Eckener produced a short tract entitled "Brief Instructions and Practical Hints for Piloting Zeppelin Airships." Although this document soon became the bible for airship commanders, parts of it would have made alarming reading for prospective zeppelin *passengers* had they ever become acquainted with its mix of general precepts, caveats, warnings, platitudes, rules of thumb, formulas and nostrums, and advice for coping with unexpected situations, such as, "Should the ship go out of control upwards or downwards, it is recommended first of all that the motors be set on half speed."

Liberally peppered with exclamation points, Eckener's book was divided logically into sections entitled "The Takeoff," "The Airship Under Way," "Landing Under Various Conditions," and "Lying Out in a Storm and at Anchor," with subsections covering "Engine Trouble," "Casualties to Cells," "Icing," and other such threats aloft. Judging from Eckener's flight manual, piloting a zeppelin required constant vigilance and was not for the faint of heart, nor for anyone with weak nerves.

"Keep your eyes open!" he wrote regarding the takeoff procedure. "Stand by to drop ballast forward or from amidships, if the bow or the entire ship will not climb!"

He advised taking off as soon as possible after removing the ship from the shed, for there was nothing to be gained by pointlessly hanging around the airfield. "In gusty weather the ship yaws and plunges; with fine weather the gas warms up and begins to blow off. . . . If it is raining, one should make special haste to get away."

The commander had to pay extra-close attention during the first hours of a flight, for "at any moment a malfunction or even a breakdown may occur in the cells, valves, ballast sacks and emergency ballast sacks, and in the rudder, ballast and engine telegraph wires."

In cruising flight, the commander should fly neither too high, because higher altitude meant less lift, nor too fast, for the faster the ship went the greater was the tendency to pitch up and down.

"One must always consider the possibility of gondola fires, of explosions in the mufflers, and of electrical discharges on a ship's hull, even in cloud formations which appear harmless."

Zeppelins had two helmsmen, one each for the elevator and rudder wheels, which were located at the port side and bow of the control gondola, respectively. The rudder man controlled heading (as indicated by the compass), the elevator operator controlled the airship's nose-up or nose-down attitude (as shown by the inclinometer). The elevator man had to be particularly sensitive to control wheel forces, since a need to make constant corrections could reveal "serious and even critical static phenomena within the ship, for example the emptying of a gas cell. . . . It is therefore urgently necessary that the commander and the elevator man should follow attentively the ship's dynamic performance at all times."

All this applied to flying in calm, stable, fine weather. In *bad* weather, flying a zeppelin approximated a state of continuous de facto emergency. Flying through rain could make the ship heavier by a ton or more, whereas "cloudbursts during thunderstorms can momentarily load the ship to still higher values, whose effect will be further reinforced through the dynamic effect of falling water."

Entering a thunderstorm was a definite no-no, given the possibility of lightning, which could set fire to released hydrogen gas. "Every equalizing spark between ship and cloud, which does not always have to take the form of real lightning, can, if the ship is valving gas, easily discharge into a layer of explosive gas and lead to ignition. Also the possibility cannot be excluded that lightning leaping from cloud to cloud may accidently strike through the stream of valved gas and set it afire."

There were additional warnings and strategies for dealing with all manner of problems, from damaged cells, stuck valves, and ice accumulations on surfaces to flying blind through fog or cloud banks—after regaining one's bearings when once again in the clear, Eckener advised, "follow first-class railroads or highways."

A few of Eckener's "practical hints" were eerily prescient of the situation that would be faced by the *Hindenburg*'s commander some eighteen years later. During the landing procedure, said Eckener, it was especially important to keep the ship in balance by the adroit release of water ballast from the nose or tail. But if the cumulative effect of doing so was not enough to manage the craft, the captain was encouraged to order crew members to run fore or aft along the axial or keel corridors, like a herd of trained hamsters, in a further attempt to balance the ship. (In airship parlance, crew members engaged in such human-ballast exercises were referred to as "galloping kilos.") Both of these tactics, and more, would be tried on May 6, 1937, by the commander of the *Hindenburg*, Max Pruss.

Eckener seemed to have anticipated every possible worst-case scenario and had proposed one or more practical methods for dealing with them. His parting words of wisdom: "In general one should remember: keep cool!"

Chapter 6

A TECHNOLOGICAL ANOMALY

Of all the technologies or technological artifacts that have ever been invented throughout history, very few of them are inherently harmful or destructive. Such crackpot medical procedures as bloodletting, while actively damaging to a patient's health, do not quite rise to the grand level of a technology. Still, many beneficial technologies have been thought to be evil, crazy, or blasphemous ("playing God") or to have other objectionable characteristics at the time they were invented. The list of technologies or artifacts that have been prohibited or banned is a long one and includes crossbows, guns, mines, nuclear bombs, electricity, automobiles, large sailing ships, bathtubs, blood transfusions, vaccines, television, computers, and the Internet, among other things. Indeed, even some of today's most powerful technologies are controversial to the point that critics have proposed outright bans on the use or further development of such innovations as genetic engineering, nanotechnology, and synthetic biology. Various governments have actually banned a number of recent technolog-ical artifacts: the BlackBerry (banned in the United Arab Emir-ates and Saudi Arabia), high-power laser pointers (Australia and many countries in Europe), and the iPad (banned for a short time by Israel). As for the United States, our own government has at one time or another tried to prohibit the introduction and use of several technologies or machines, among them videocassette recorders, genetically modified foods, embryonic stem cells, the phonograph, and margarine.

Such prohibitions are usually short-lived. On the other hand, a small subset of technological systems or objects are genuinely pathological in the sense that they are so inherently flawed by be-ing too risky, not worth their cost in human lives (much less dol-lars or resources), and unacceptable for other reasons that it was foolish to have developed them in the first place and would be inadvisable to develop them further. The hydrogen airship as used for passenger travel is arguably one such pathological technology. This point was first realized, not by the Germans, but by the Brit-ish. The British, it turns out, ran an airship program of their own for twenty years, and as bad as the German experience with zep-pelins often was, that of the British was even worse. Far worse.

That the British had any airship program at all is strikingly odd to begin with, for several reasons. The first is that the British were well acquainted with the dismal German record of crashes and di-sasters but proceeded with their own airship project anyway. The second is that in England neither the general public nor those in the military or the government ever experienced any sort of zeppe-lin madness or Delirium. Seeing a zeppelin in flight did not cause mass hysteria or prompt the average Briton to burst into tears or into spontaneous and impassioned renditions of "God Save the King" (the British sovereign at that time being King George V).

Far from regarding airships as sublime or transcendental objects, during the Great War the British grew to hate the things, and Londoners looked upon the flaming, falling destruction of a zeppelin as a special form of high entertainment, oftentimes even applauding as it plummeted to the earth, burning brightly. The British press, meanwhile, portrayed zeppelins as lumbering, bumbling gas bags of no particular military, strategic, or other value. Winston Churchill, who was for a time First Lord of the Admiralty, once reflected that "I rated the Zeppelin much lower as a weapon of war than almost anyone else. I believed that this enormous bladder of combustible and explosive gas would prove to be easily destructible." As in fact it had been.

Finally, the British had no avuncular, grandfatherly cult figure whom they regarded as a hero or inventive genius in the way that the Germans viewed Count Ferdinand von Zeppelin, and so they were not motivated to build zeppelins for sentimental, cultural, symbolic, or patriotic reasons. Why then did they nevertheless go on to design, build, fly, crash, burn, and die in their own home-built bladders of combustible and explosive gas?

According to zeppelin historian Douglas Robinson, two reasons were simple curiosity and envy on the part of the British navy, which wanted to be on an equal footing in the air with its German counterpart (just as Zeppelin wanted Germany to be on a level with the French). The Royal Navy had no desire to be seen as backward or uncompetitive, and perhaps its officers even imagined that, by learning from the mistakes made by the Germans, they could beat the zeppelin-makers at their own game. More specifically, a number of Royal Navy officers thought that airships would be useful as long-range aerial scouts in the North Sea and over the Atlantic approaches to England. As a result, on May 7, 1909, the British Admiralty placed an order with Vickers Ltd., a company that normally built submarines, warships, and other ordnance, for His Majesty's Airship No. 1, which would be more popularly known as the *Mayfly*.

The *Mayfly* would be 512 feet long by 48 feet in diameter, slightly larger than the German zeppelins of the era, and would be built on the same type of longitudinal girder and transverse ring internal framework. It would even contain the usual seventeen gas bags made out of rubberized fabric. The craft was designed to take off from and land on the water, and so it was outfitted with such nautical gear as an anchor, capstan, and hawsers.

After a construction period of two years, the builders belatedly discovered, upon its completion in May 1911, that the ship was so overweight it would not lift off from the water. It merely sat there complacently on the surface of Morecambe Bay, in north-west England, as if it were a proper seagoing vessel. A force of 300 sailors pushed the *Mayfly* back into its shed, where workers undertook several weight-loss procedures, including the removal of its keel. This was like eliminating the foundation just as the house was complete.

The next time the craft came out of the shed, in September, a series of ominous cracking sounds emerged from amidships, whereupon the *Mayfly* broke neatly in two, as if chopped in half by a meat cleaver. The two halves rose briefly into the form of a "V," and crew members dove from their gondolas and swam for their lives, after which both sections of the divided ship splashed back down into the bay.

Churchill, after this exhibition, observed that the ship would have been more accurately called "the *Won't Fly*." Rear Admiral Sir Doveton Sturdee, upon viewing the floating two-part wreckage, remarked: "The work of an idiot!"

The craft was put back into its shed, where it was left to rot. And that was that.

After those humble beginnings, the British airship program had nowhere to go but up, which it did with the next rigid airship, the No. 9. (British airships seem to have been numbered essentially at random.) This one actually flew, but it lasted for a total of less than 200 hours before it too was damaged and dismantled. On August 16, 1918, another (the R 27) was accidentally set afire in its hangar (by an American naval blimp crew, no less), where it burned to the ground, thereby becoming *Hindenburg* prequel number 7. Others fell apart and were soon decommissioned, broken up, and forgotten.

The subsequent history of the British airship program amounts to a Greek tragedy, complete with momentary triumphs of the hero that give an illusion of success before the emergence of a fatal flaw, together with the misfortunes that stem from it, heralding his final exit from the stage.

One of the triumphs was HM Airship R 34, which in July 1919 flew across the Atlantic to Roosevelt Field, Long Island (the airport from which Charles Lindbergh would set out for Paris almost eight years later), then returned to England without fatal mishap. It was the first round-trip crossing of the Atlantic by air. That was one of two high points of the British experience with airships, but it was followed in due course by a low point when, in January 1921, the very same ship (R 34) ran into a hill at night, damaging the control car and two engines. The craft limped back to a mooring mast at Howden, an airship base in Yorkshire, where the wind effectively blew it to bits.

An even lower point, but hardly the nadir, was reached later that year when a follow-on ship, HM Airship R 38, was being prepared to embark on its own trip to America. Since there were questions about the craft's structural integrity (some transverse girders having failed on its third test flight in July 1921; they were later repaired), someone in a position of authority (it was never

clear who) ordered that the R 38 be subjected to a series of in-
creasingly severe control inputs, as a kind of aeronautical stress
test. The theory was, apparently, that if the ship could survive
these induced, violent maneuvers, then it would be safe to fly it
across the Atlantic.

And so, during its next test flight, on August 24, 1921, with the
ship traveling at high speed, the rudder man started in with the
process of putting the helm over to port, then to starboard, and
back again, at progressively sharper angles each time. The ship
was 699 feet long, and the top and bottom rudders being slammed
back and forth produced ever greater bending moments and yaw
forces on the hull. Finally, the longitudinals toward the rear of the
craft gave way, and the ship separated into two sections, as if in a
Mayfly rerun. Nobody had been killed in the *Mayfly*, but when the
R 38's forward section fell into the River Humber, near the town
of Hull in northeast England, some of its gas cells exploded, shattering windowpanes in the immediate area, and then the entire
front end of the ship burst into flames. The fire killed forty-four
of the forty-eight men onboard, which was eight more than would
die in the *Hindenburg*. The R 38 thus became *Hindenburg* prequel number 8.

That disaster should have ended the British program, but it
didn't. What kept it going was the government's desire to connect
the far-flung outposts of empire by air. As it was, travel between
London and places like India, Australia, New Zealand, and South
Africa took weeks by steamer, whereas an airship could make the
same voyages in a matter of days. In 1924 the British government
put forward a plan known as the Imperial Airship Programme,
whereby two gigantic new airships would be built, one by private enterprise, the other by the government itself, although both
projects would be financed by the state. The Airship Guarantee

Company, a subsidiary of Vickers Ltd., a private corporation, would make the R 100 (called the "capitalist ship") at the former air station at Howden, while the British Royal Airship Works at Cardington, in Bedfordshire, would manufacture the R 101 (the "socialist ship"). Both of them were to be built to the same general specifications: each would be able to carry 100 passengers at a cruising speed of at least sixty-three miles per hour.

The idea behind this capitalist versus socialist scheme was that a head-to-head competition between the two projects would further the interests of airship development: in the technological equivalent of Darwinian survival of the fittest, future specimens could combine the best features of both designs. But in the end the competition would in fact lead to the British airship's quick, proper, and permanent extinction.

As opposed to the "empirical," or trial-and-error, approach to design as routinely practiced by Count Zeppelin and his engineer Ludwig Dürr, both of the British ships would be products of substantial amounts of theorizing, calculation, and technological advance planning. The chief "calculator" (engineer) on the capitalist ship, responsible for estimating stresses on the R 100's girder system, was one N. S. Norway. Better known under his pen name Nevil Shute, Norway was the author of the novel (and later film) *On the Beach*. In his autobiography, *Slide Rule,* Shute described the construction, proving flights, and ultimate fates of both ships.

Working on the capitalist R 100 was an often harrowing experience for Shute, who not only had a fear of heights but also harbored a particular distaste for the engine trials that were held just prior to the ship's completion. These were conducted inside the construction shed, where only fifteen inches of clearance separated the floor and the tips of the big wooden propellers, on a hull that contained five million cubic feet of hydrogen. "Whatever precautions we took it was impossible to keep the hull of the ship from surging up and down in the fierce air currents generated by the thrust of the propellers," he wrote. "If a propeller had hit the floor or if a suspension cable had parted under that test the issue could only have been sheer disaster and the loss of many lives."

He expected no disasters in cruising flight, but was less confident about landings, during which a ship could hit the ground in a manner that would cause broken electrical connections that could generate sparks. "In such a case the only chance for survival would be to jump on to the inside of the outer cover and cut one's way out and drop down to the ground, and one would have not more than five seconds to do it in. For this reason before the first flight of R100 I bought a very large clasp knife and sharpened it to a fine point and a razor edge, and I carried this knife unostentatiously in my pocket throughout the flights that the ship made."

One of those flights, in August 1930, was to Canada and back. The ship encountered the usual number of airship problems en route: one of the gas cells leaked and steadily lost hydrogen and lift, and the fabric on some of the control surfaces was torn away by the slipstream, in one case creating "a hole large enough to drive a bus through." In addition, the lights went out; part of a propeller flew off and penetrated the hull; an engine failed; and while flying through a thunderstorm the craft at one point assumed a thirty-five-degree nose-down angle, forcing the crew to hold on tightly to keep from falling.

The Canada flight, Shute thought in retrospect, was premature. Nevertheless, "we did it, and got away with it."

When the "socialist" R 101 was complete, tests showed that it was grossly overweight despite all the calculations, theorizing, and wind tunnel tests done beforehand. The builders removed various items to reduce its mass by three tons, and they also removed some of the gas-bag restraining wires so as to increase the volume of the cells, and hence their lift, by *another* three tons. But with the restraining wires gone, the cells rubbed against the hull framework, and the chafing created hundreds of holes in the fabric, a situation that the builders dealt with by installing cotton padding at the points where the abrasion occurred.

The loss of the cell restraining wires had the further, more serious drawback that it permitted the cells to sway back and forth inside the ship by as much as fourteen feet. This movement of their centers of lift made the ship dynamically unstable and set up conditions for a vicious cycle of cause and effect: when the cells swung forward they made the nose rise up, which let the cells move even farther forward, and so on, yielding an undulating, porpoising flight path through the air that the helmsman had to counteract by constantly making a series of opposite and precisely timed control inputs.

But even after the weight reduction of six tons, the ship was still too heavy, and at this point, according to British airship historian J. A. Sinclair, "R 101 should have been scrapped." Instead, to make for yet more lift, the government builders at the Royal Airship Works took the extreme measure of simply cutting the ship in half and adding in another cell. It was now a huge ship, 777 feet in length, just 27 feet shorter than the *Hindenburg* would be. The extra cell increased the craft's lift to a more acceptable level, but it also altered its balance and dynamic performance in ways that had not been foreseen, precisely calculated, or planned for in advance. Despite the risk of flying off with these known unknowns still unknown, the time had come for the R 101 to make *its* premature proving flight, which in this case was all the way to India.

The rush into the air on this occasion was prompted by politics, in the person of the British air minister, Lord Christopher Birdwood Thomson, a man who had himself been born in India and now harbored dreams of becoming the colony's next viceroy.

Thomson, like Count Zeppelin, had been an army officer and had no particular credentials in aeronautics, although he was a rabid fan and supporter of the airship, which he regarded with an almost messianic fervor. In the annals of airship lore, Thomson is remembered for making two statements in the famous-last-words tradition. One was delivered before the House of Lords in June 1930: "This is one of the most scientific experiments that man has ever attempted. There is going to be no risk while I am in charge. No lives will be sacrificed through lack of foresight and skill." The other was made shortly before he boarded the R 101 for its maiden flight to India: "She is as safe as a house—except for the millionth chance."

The ship was flying to India with Thomson aboard so that he could travel there and back in time for an Imperial Conference to be held in London on October 20, 1930. As Nevil Shute put it, the R 101 "would arrive back in England on October 18th in order that Lord Thomson could step into the conference room as fresh as a daisy on October 20th."

And so, on October 4, 1930, the R 101, at that time the largest airship in the world, newly stretched and filled nearly to bursting with more than five million cubic feet of hydrogen gas, lifted off from its mooring tower at the Royal Airship Works at Cardington and headed south toward the English Channel in the dark of night. All available indications were against it: the ship was so overweight at takeoff that it was necessary to drop four tons of water ballast; the holes in the gas cells had been patched, but the cells were still swaying and rubbing against the cotton-padded framework;

headwinds of up to fifty miles per hour had been forecast; one engine failed shortly into the flight; and when the craft encountered heavy, driving rain, low clouds, and waves of turbulence, its innate and already disturbing instability was only amplified.

In the wee hours of October 5, just after 2:00 a.m., the ship suddenly nosed down, descended, and ran into a hillside near Beauvais, France. The impact was followed by an explosion and a violent hydrogen fire that completely destroyed the craft. Lord Thomson, along with forty-seven others, died.

At that point, finally, the British had gotten the message: this was not a technology worth pursuing; rather, it was a technology very much worth abandoning. So they got out of the airship business once and for all. Their sole remaining ship, the capitalist R 100, was broken up and sold for scrap.

Still, it was private enterprise that was the winner in this somewhat peculiar competition. The "socialist" ship, the R 101, now became *Hindenburg* prequel number 9.

It might be difficult to believe, in light of all this, that any nation could possibly have an even worse experience with an airship program than the British had with theirs, but one nation in fact did. And that was the United States. If there were a booby prize in the category of airship failure rates, accidents, catastrophes, and deaths, the Americans would win it by a landslide. Bad as they were, the worst airship disasters in history were neither that of the British R 38, which killed forty-four, nor the R 101, which killed forty-eight, nor even the *Hindenburg*, in which thirty-six lives were lost. No, it was the American airship *Akron* that took the greatest number of human lives when it crashed off the coast of New Jersey in 1933, killing seventy-three, a death toll more than twice that of the *Hindenburg*.

America's woeful airship record was doubly ironic in light of the fact that the United States had the British and German experiences with airships to learn from and build upon, and the further fact that the Americans enjoyed an apparent advantage over their European counterparts in that it had access to, and a monopoly over, helium gas. Helium was a nonflammable lifting agent that was found mainly in the natural gas fields of Texas, and so with it filling their gas cells, American airships ought to have been considerably safer than those whose gas bags were inflated with hydrogen. But that expectation was not borne out by the actual course of events. For helium was not an unalloyed blessing, but came with its own set of drawbacks and penalties. Indeed, in some ways it was a curse. Since it was heavier than hydrogen, it provided less lift. It was also rare and therefore expensive, and in consequence airship captains went to extraordinary, excessive, and sometimes even dangerous lengths to prevent it from being vented off uselessly into the atmosphere.

The American program got started to begin with because the US Navy had inklings of a war in the Pacific and thought that an airship's extended range of action would make it a useful tool in scouting and reconnaissance missions. In 1919 Congress appropriated funds for two airships: one to be purchased from the British, the other to be built within the United States. As it turned out, neither of them was exactly a success.

The British ship in question was the R 38, the ill-fated craft that would end up by splitting apart, exploding, and then burning over the River Humber in England in August 1921. The Americans, who had contracted for the ship two years prior to the debacle, intended to use it as a training vehicle for their future airship fleet. At the time it was built, the R 38 was the largest and most advanced airship in existence: it was 699 feet long by 85 feet in diameter, with a gas volume of 2,724,000 cubic feet. Construction began in February 1919, and in April 1920, long before it was finished, a group of twenty-seven American naval officers and enlisted men arrived in England to begin training. The plan was that after several hours of flight time in existing British airships, the American crew would transition to the R 38 upon its completion and then fly it across the ocean to Lakehurst, New Jersey, where it would take up residence inside a vast new hangar that had been specially built for the purpose.

By the time it crashed, the R 38 had been painted with American colors and naval insignia and bore the US Navy's official numerical designation, ZR 2. To all intents and purposes, it was already an American airship. But by this time it was also damaged goods, since some of its transverse girders had failed during its third test flight. Although the British had replaced the broken girders and had even added some reinforcements to them, the Americans suspected that the ship was structurally weak. Nevertheless, they went ahead with plans to take delivery and bring the craft to the States.

That was yet another example of the kind of bad judgment and triumph of hope over reason that was so often associated with airship development. For as it happened, "America's first airship" never made it across the Atlantic. In fact, it never even left England. And when it broke up and burned over the River Humber, it killed sixteen American men, wiping out the majority of the US Navy's body of trained airship personnel.

The country's second airship, and the first to fly within its borders, was the ZR 1, the *Shenandoah*. The craft would be built in the United States, by the navy itself, which would have exclusive and total control over its design, material composition, and construction, and so one might reasonably expect that the ship

would be crafted to a higher standard, and to more exacting specifications, than its unfortunate British predecessor. On the other
hand, the navy had never before put together a large rigid airship,
and so the ZR 1 would not be a design de novo, or built from
scratch. Instead, the ship would be based on prior art, in this case
on the German military zeppelin L 49.

The navy chose the L 49 as a template not so much for its virtues, or because it was an ideal specimen of the type, but rather
because it was a German ship for which they happened to have
detailed plans. The blueprints for the L 49 had become available after a battle over France in October 1917, when a number
of French fighter planes forced the ship down near the town of
Bourbonne-les-Bains. It was traditional among German zeppelin
commanders during the war to set fire to zeppelins grounded in
foreign territory specifically to prevent the enemy from acquiring
structural or construction information. (One imaginative commander conceived the alternative plan of driving his craft into the
ground vertically, collapsing it up "like an accordion.")

The captain and crew of the L 49, however, were captured before they could set their ship ablaze. The French regarded this as
a golden opportunity to learn the "secrets" of the zeppelin and to
draft a written record of the ship's design, of which they later sent
copies to the United States.

But the joke was on the French (as well as on the Americans)
because the Germans had built the L 49 not for strength but
rather as a high-altitude zeppelin (a species also known as a "height
climber"), and for that reason they had intentionally made it as
light and essentially as flimsy as possible. And so in using this craft
as their model for the *Shenandoah,* the Americans were in effect
producing a ship that embodied certain inherent structural weaknesses to begin with, as well as certain operating limitations and

conditions that were bound to show up sooner or later in flight. The US Navy made a few modifications to the original German design, but whether they were actual "improvements" was questionable.

The navy fabricated parts for the ship at the Naval Aircraft Factory in Philadelphia, which until then had produced mostly fighters and bombers, and then shipped the components to Lakehurst for assembly. The first flight of the ZR 1 *Shenandoah* took place on September 4, 1923; coincidentally, this was also the world's first flight of a rigid airship filled with helium. It therefore seemed as if the ship would be exceptionally safe and secure against disaster. But ironically, the craft's apparently innocuous, nonflammable, and "safe" lifting agent may have played a large role in its subsequent, and bizarre, undoing.

Helium was first discovered in 1868, not on earth but on the sun. The French astronomer Pierre Jules Janssen was in India observing a solar eclipse when he detected a bright yellow line in the solar spectrum, a spectral signature that did not appear to belong to any known element on earth. (Later, in 1870, Janssen would escape the Siege of Paris by balloon. And to complete this triple-header of lighter-than-air milestones, in 1885 he would also photograph from his Paris observatory the Krebs and Renard dirigible, *La France,* in flight.) Janssen sent a report of his discovery to the English astronomer Norman Lockyer, an expert on solar spectra. Lockyer couldn't identify the spectral line either and hypothesized that it was the unique signature of a new element that he called helium, after the Greek word for the sun, *helios.* Helium was not actually isolated on earth until 1895, by the Scottish chemist William Ramsay, among others.

Although it is the second most abundant element in the visible universe (the first being hydrogen), helium is extremely rare on earth—so rare that during the 1920s helium gas cost $120 per thousand cubic feet at a time when hydrogen was selling for $2 to $3 for the same amount. And so, to preserve its supply of this precious substance, the *Shenandoah*'s captain, Zachary Lansdowne, ordered that ten of the craft's automatic pressure relief valves be removed from the ship, the better to prevent their otherwise continuous and inevitable leakage of small-molecule helium from its cells. The downside of this modification, however, was that if the ship were ever caught in a violent updraft, such as are often found in thunderstorms, the cells might be unable to release their gas fast enough to prevent them from bursting.

The *Shenandoah* encountered just such a storm over Ava, Ohio, on September 3, 1925, when a rising current of air lifted the ship from an altitude of 1,600 feet to more than 6,000 feet in a matter of minutes. There now occurred one of the most extraordinary sequence of events ever to befall any airship: the structure broke up first into two halves that were temporarily held together by the rudder and elevator cables, which finally snapped, leaving the nose section divided from the stern section. But then each of *those* sections in turn broke up into two pieces: the gondola detaching from the nose portion, sending Zachary Lansdowne and six other crew members to their deaths. Then the stern section itself broke up into two, and the separate fragments both fell to the earth.

That left the remaining nose section still aloft, with seven men inside it, including the navigator, Charles E. Rosendahl (who would be the commander of the Lakehurst Naval Air Station at the time of the *Hindenburg* disaster). They got caught in another updraft that carried them up to 10,000 feet. Enough of the bow's internal components had remained intact, however, that by deliberately slashing the gas bags and carefully dropping ballast,

Rosendahl over the space of an hour managed to bring the nose section to a reasonably soft landing. Incredibly, he and most of the others walked away from the wreckage unhurt.

A total of fourteen men died in the accident, in which there were twenty-nine survivors. Because of the manner in which the craft broke up, with the various portions being blown across the landscape by the wind, the four principal parts of the *Shenandoah* came down at three separate crash sites, in Buffalo, Noble, and Sharon Townships in Ohio. Each of the crash sites is today marked by a small memorial.

Although the *Shenandoah* had essentially disintegrated mid-air into isolated bits and pieces, that airship was luckier than the next one to be constructed in the United States. The *Akron*, ZRS 4, built at the "Airdock" of the Goodyear Tire and Rubber Company, in Akron, Ohio, was accident-prone from the start and repeatedly escaped the control of its ground handlers, even when they numbered in the hundreds. It also managed to detach itself from a 133-ton beam that had been especially developed for the purpose of holding the ship securely to the ground. Finally, on April 4, 1933, the *Akron* became the deadliest airship in history when it flew into a thunderstorm near the town of Barnegat Light, New Jersey, fell into the sea, and sank, killing all but three of the seventy-six aboard. A navy blimp sent out to look for survivors also crashed, killing two additional men, making for a combined total of seventy-five deaths.

Almost two years later, in February 1935, the *Akron*'s sister ship, the *Macon* (ZRS 5), terminated the nation's flirtation with airships when it was hit by a gust and crashed into the Pacific off Point Sur, California. Only two out of eighty-three aboard died, but by this time it didn't much matter. The American public, Congress, and even the US Navy itself were sick of the big military rigids.

It was an irony of history that of the country's five airships, the only one that did not fall apart, break up, burn up, explode, or otherwise meet with disaster and death was one made by the Germans: the Zeppelin LZ 126, US Naval designation ZR 3, also known as the *Los Angeles,* which was built under contract for the US Navy by the Zeppelin Company in Friedrichshafen. The *Los Angeles* had its share of problems—engine failures, leaking gas cells, weak girders—but it never killed a single person, and it did not suddenly disintegrate in flight. It was the sole American airship to have a peaceful ending when it was dismantled at Lakehurst in 1939.

In view of this long, steady, and multinational trail of airship catastrophes, the question arises as to why these vehicles were so technologically pathological across the full span of their operational lifetimes—short as that was in many cases. To this there are a number of answers. Hydrogen was an obvious explanatory factor in those airships containing it, which comprise the vast majority of all zeppelins ever built. Such ships were vulnerable to sources of ignition from both inside and outside the craft: inside from broken electrical connections, from static electricity generated by friction between gas cells, from snapped cables; outside from lightning, St. Elmo's Fire, and other sources of atmospheric electrical discharge. Whatever the cause, exposure to such phenomena presented an entirely unacceptable level of risk for civilian passenger travel, one that would not be tolerated today by any country's regulatory agency governing the safety of civil aviation. In fact, today the carriage of paying passengers aboard such a craft would be regarded as an act of wanton criminal negligence, if not sheer lunacy.

But there is a more basic explanation for the pathological nature of the rigid airship: namely, its size, in particular the extreme disparity between the scale of the craft and the size and scale of those who intended to travel in it, and above all to control it. Any ship that routinely needs a force of 300 to 500 human beings to stabilize it during exits from and entrances to the hangar, as well as in takeoff and landing maneuvers, and is *still* often largely uncontrollable anyway, is a machine too big for its makers and supposed masters. Airships were repeatedly torn away from their mooring masts, crushed against hangar walls, and yanked by the wind from the hands of their ground crews.

In February 1932, a year before it plummeted into the ocean and sank, the airship *Akron* was scheduled to make a safety demonstration flight at Lakehurst in front of a group of skeptical congressmen. The *Akron* weathervaned into the wind, with the tail end heading in the direction of the spectators, who now scattered in all directions. The rear of the craft dropped to the ground, crushing the vertical fin, despite the fact that the craft was releasing water ballast from the stern, like a giant beast urinating in flight. All of this havoc was caught on film (which can be seen on YouTube).

Three months later, with the fin rebuilt, the *Akron* flew to California. While attempting to land at Camp Kearney, ground crew members grabbed onto the landing lines as usual, in an attempt to haul it down out of the sky. But the ship refused to be hauled down; instead, it rose back up into the air as the crew members who were still holding on dropped off the landing lines, one by one. Two of the men fell to their deaths—an event also captured on film (and on YouTube).

The size of an airship made it extraordinarily susceptible to wind, both when aloft and near the ground. A zeppelin was like an immense tubular kite or sail, and, being filled with practically nothing, while in many cases also being underpowered, the craft was all too often hostage to the elements.

Turbulence, either of the clear-air variety or the kind associated with thunderstorms, damaged, dismembered, or destroyed numerous zeppelins by subjecting their comparatively frail, skeleton-like airframes to unsustainable and intolerable physical stresses: bending moments, twisting forces, and expansions and compressions. Nobody would tolerate a seagoing vessel that could not survive rough seas, but rigid airships for the most part could not survive rough weather. That fact was routinely minimized, ignored, or pushed under the rug by dyed-in-the-wool airship advocates, enthusiasts, and spellbound apologists.

The size of the craft also made it slow to respond to control inputs, whether of throttle, rudder, or elevator, the valving of gas, or the release of ballast, making for conditions that further put passengers, crew members, and ground handlers at risk.

All of these numerous deficiencies, incongruities, and engineering stupidities gave these expensive craft exceptionally short life spans, especially as compared to the extended service lifetimes enjoyed by conventional aircraft. In 1924 a list of zeppelins compiled by the British science journal *Nature* showed that the "average expectation of life of a Zeppelin falls far short of eighteen voyages in eighteen months." By contrast, the Douglas DC-3, a propeller-driven civil airline classic that made its inaugural flight in December 1935, the same year the *Macon* crashed, is still in active commercial service today, eighty years after its introduction.

The zeppelin was pathological in another sense as well. Some specimens flew for extended periods with no major accidents, killed no one, and traveled everywhere, thereby fostering the illusion that hydrogen airships were basically "safe." The greatest illusionist of them all was the ship named after its inventor, the *Graf Zeppelin*. The *Graf* made a round-the-world trip, went to the North Pole and back, established a regularly scheduled passenger service to South America, crossed the North Atlantic many times, and ended up carrying more than 13,000 passengers over a total distance of more than a million miles (the first aircraft in history ever to do so), on almost 600 flights, without killing a soul. Furthermore, the *Graf* did all this with incredible sangfroid, relative impunity to bad handling, and even, on occasion, recklessness. It did everything but travel to the outer planets, their moons, and back, and to top it off, it lived to a ripe old age (of nine, absolutely ancient for a zeppelin) before being retired from active service in 1937 and finally dismantled in 1940.

So very lucky was the *Graf* that it appeared to be mystically exempt from the otherwise inviolable laws of airship nature. The next ship to come off the Zeppelin Company's assembly line would have no such good fortune. It was the *Hindenburg*.

Chapter 7

DEATH RATTLE
OF A LEVIATHAN

The *Hindenburg* was the 118th airship to be built by the Zeppelin Company, and Hugo Eckener considered the craft to be "the ideal Zeppelin."

The story of the *Hindenburg* is full of ironies, stupidities, missed opportunities, and tragedies, one of them being that it very nearly escaped its ultimate fate. The fact is that when the craft was conceived, it had been designed for helium, not hydrogen. Only later, after it had been redesigned as a hydrogen ship, did its final destruction seem, in retrospect, predetermined.

For the Zeppelin Company, the decision to fill the new craft with helium represented a major change in corporate policy. Helium was inordinately expensive, about seven times the cost of hydrogen. It also provided less lift, meaning that a ship filled with helium would be able to carry fewer passengers and less cargo and mail than a hydrogen ship of the same size. But there was an even more daunting obstacle to the use of helium: Germany had next to no supplies of the gas, while the United States had a virtual monopoly on the stuff. And under the US Helium Control Act of 1927, sales of helium to foreign countries was prohibited by law.

So in the face of all this, why then had the *Hindenburg* nevertheless been intended for helium? Because just as its design was getting under way, the British R 101 "socialist" airship had crashed, exploded, and burned, killing forty-eight of its fifty-five passengers and crew and strewing a spectacularly ugly wreckage across a hillside near the coast of France. That had happened on October 5, 1930. The first design memorandum pertaining to what would become the *Hindenburg* which was then known within the Zeppelin Company only by its "works number," *Projekt LZ 129,* was dated December 29, 1930, barely three months later.

If all had gone according to the rosiest possible scenario, Hugo Eckener, the Zeppelin Company's president, would have applied to the US government for a waiver of its helium export prohibition, the government would have granted his request, the Germans would have gotten their helium, and the *Hindenburg* disaster never would have occurred. Apparently, however, Eckener never made any such request to begin with. To confirm that he in fact did not, in 1990 the American zeppelin historian Henry Cord Meyer made an exhaustive search of the available relevant documents, looking for evidence that Eckener or someone else had applied for a helium export waiver. He searched through the Zeppelin Company archives at Friedrichshafen; the records of the Goodyear Corporation in Akron, Ohio; the US National Archives in Washington, DC; and the archives of the German Foreign Offices in Bonn and in Koblenz. At the end of his search he reported that "I have not found one specific piece of evidence that the Germans or Germany ever specifically applied to the US government to inflate LZ 129"—that is, to inflate it with helium.

After the *Hindenburg*'s destruction at Lakehurst, Eckener, at a US House of Representatives committee hearing, offered an explanation for why he never made a formal request for helium: "At that time all necessary facilities for transportation, storage, repurification, and so on, were not available."

Taken by itself, that was a fairly lame explanation. If he had really been serious about obtaining helium for the *Hindenburg*, the obvious thing for Eckener to have done would have been simply to construct the necessary facilities, or to make arrangements to have them erected. After all, the company had already built an enormous new construction shed at Friedrichshafen, weighing in at 1,100 tons, to house the *Hindenburg* before it built the new ship itself.

That Eckener made no effort to fabricate the necessary infrastructure suggests that other reasons were actually at work. Airship historians have often pointed out that German zeppelins had been operated commercially for years without a single passenger fatality. And the safety record of the German passenger zeppelins may have given the *Hindenburg*'s builders the false impression that they could operate the new craft at minimal risk even without helium.

It was equally true, however, that between 1908 and 1930 no fewer than twenty-six hydrogen airships had been destroyed by fire due to accidental causes, not even including those shot down in wartime, killing a total of 250 people. In other words, there had been twenty-six separate *Hindenburg* prequels, with 250 deaths, yet zeppelin enthusiasts incomprehensibly continued to insist that hydrogen airships were "safe." It appears, therefore, that the *Hindenburg*'s builders failed to use helium through a combination of bad judgment, hubris, and hydrogen addiction.

The true reasons why helium was not used in the *Hindenburg* might never be known with certainty. In any case, this was clearly an opportunity missed. In the end, LZ 129 had been redesigned for, built for, and inflated with hydrogen. Ever since the disaster at Lakehurst, people have been looking for omens that presaged the tragedy. The rejection of helium must surely rank as ill portent number 1. There would be others.

A part from its fiery ending, what the *Hindenburg* had always been known for was its size. It and its later sister ship, the *Graf Zeppelin II*, remain the largest aircraft ever built. Of the many attempts that have been made to convey just how big it actually was, perhaps the most successful comes from Airships.net historian Dan Grossman, who has pointed out that the *Hindenburg*, from nose to tail, was longer than the United States Capitol Building. The Capitol Building, including both the House and Senate chambers that flank the main rotunda, spans a distance of 751 feet, 4 inches. The *Hindenburg*, at 803.8 feet, was more than 50 feet longer than that. It was almost one-sixth of a mile long, roughly as long as a New York City block between avenues. It is understating the case to say that these stupendous dimensions gave the craft an undeniable physical presence that transfixed the mind and senses and cast a hypnotic spell upon virtually all who laid eyes on it.

Vertically, the distance from the top of its upper tail fin to the bottom edge of its lower fin was 150 feet, roughly equivalent to the height of a 13-story building. The fins themselves were about 100 feet long, and they were so spacious inside that the lower vertical stabilizer included an alternate control room equipped with "emergency helms" in the form of duplicate elevator and rudder control wheels, a minimum of instrumentation, and space enough for two or three people to move around in.

The craft's inner framework consisted of 36 triangular duralumin girders running horizontally, which collectively worked out to a total length of 5.48 miles, plus 15 vertical main rings that formed compartments for the craft's 16 gas cells, a triangular keel and walkway at the bottom of the hull, and an axial corridor that passed through the center of the ship. Ladders inside vertical shafts connected the keel and axial corridors at three different points along the ship's length. Crew members, as well as passengers on guided tours of the vessel, could easily walk through either of the horizontal corridors (although the axial corridor was more frequently used by the ship's riggers to inspect gas-cell pressure, gas valves, and the like). The interior space of the airship was additionally crisscrossed by a mazeworks of internal bracing cables and netting: the gas cells lifted against the netting, which was connected to the framework by ropes, and the ropes in turn lifted the ship. Between the cells were gas vents, maneuvering valves, and pressure relief ports, plus access ladders, ventilation shafts, rooftop ventilation hoods, hatches, platforms, observation windows, and other structures.

There were no solid beams, girders, or other structural members anywhere on the ship, except in the passenger areas, where exposed beams were left whole for aesthetic reasons. Otherwise, to achieve maximum lightness, every metal component was perforated, almost filigreed, like fine lacework, and any strut, brace, angle, channel, spar, or other supporting element had been lightened by means of stamped spacers or punched holes, as if made of Swiss cheese. The walls between the passenger cabins—"roomettes," really— were notoriously thin, allowing guests to eavesdrop on next-door conversations, arguments, or other activities.

Airship LZ 129 had a lifting gas volume almost double that of its predecessor, the *Graf Zeppelin*, at 7,062,000 cubic feet of hydrogen, parceled out in the sixteen cells, which were numbered in reverse order from stern to bow, as were the rings. The number of a given individual ring represented its distance, in meters, from an imaginary point in space just beyond the stern of the ship. The gas cells, likewise, were numbered consecutively from the tail, beginning with cell 1 and ending with cell 16 in the bow section. The cells were not lined with the usual, and expensive, goldbeater's skin, but were made of a gastight gelatin latex film (similar to photographic film) sandwiched between two layers of cotton fabric.

At ring 92 (301 feet from the rearmost portion of the ship) and farther forward at ring 140, there were lateral walkways that extended across the lower perimeter of the ship from the keel corridor to both the port and starboard sides. These lateral catwalks, made largely of wooden planks with traction rungs across them, ended at hatches that, when opened, permitted access to the engine gondolas, two at the rear and two more farther forward. The gondolas were streamlined, egg-shaped enclosures, and each contained a Daimler-Benz sixteen-cylinder diesel engine consisting of two banks of eight cylinders set in a fifty-degree Vee, an arrangement that yielded a maximum output of 1,000 horsepower apiece. Each engine drove a four-bladed, fixed-pitch, wooden pusher propeller that itself had a diameter of nineteen feet, nine inches— that is, the propellers alone were almost the height of a two-story building.

To allow for propeller clearance, the engine cars were positioned approximately ten feet away from the hull. Because each power plant was attended by a mechanic at all times during flight, a minimum of fifteen mechanics were required aboard the ship on every voyage of the *Hindenburg*. These were men who braved continuous loud engine noise (even though they wore leather flight helmets and earplugs), constant vibration, and a white-knuckle trip through the elements merely to get to their place of business. Getting from the interior of the ship to the engine car meant traversing an open, railed bridge in the case of the forward cars, or descending a ladder to the aft power cars (which were positioned beneath those in the front in order to avoid their prop-wash), all the while leaning against an eighty-mile-an-hour slip-stream and tolerating the somewhat alarming views of the passing landscape or the ocean waves below.

Beyond these strictly structural and mechanical features, the *Hindenburg* incorporated several other functional systems, mod-ules, and areas, including diesel-powered generators for lighting; fuel and ballast systems; ventilation and heating systems; drink-ing water, wastewater, and plumbing systems; steering apparatus; cargo areas; a control gondola that included a navigation room, rudder and elevator control wheels, engine telegraphs, gas-cell and ballast control boards, and flight instrumentation; passenger and crew accommodations as well as eating, drinking, and recre-ation areas; a baggage room; a post office; and an "electric kitchen" equipped with electrically powered stoves and ovens, a refrigera-tor, and an ice machine. (As a safety measure, no electrical wiring was installed above the ship's equator, and all the cabling was well insulated and enclosed within aluminum conduits.)

All of these structures and systems were enveloped by the craft's smooth outer skin, an expanse of cotton fabric that covered an exterior surface area of more than eight acres. Although the fabric was doped to make it waterproof (and also to enhance its radiance in sunlight), rainwater could increase the weight of the ship by six to eight tons.

The construction of LZ 129 began in 1932, although the fab-rication of some component parts had started the year before. Among its miles of duralumin longitudinal girders were some 11,000 pounds of the metal that had been salvaged from the re-mains of the doomed "socialist" airship R 101, which the Zeppelin

Company had purchased from the British. Supposedly, rising out of the ashes of a previous crashed airship was another bad omen for the *Hindenburg*. Still, by this stage of the game making new zeppelins out of material recovered from prior zeppelin wrecks was a well-established and accepted practice.

During the five years it took to build the LZ 129, the world—most especially Germany, and even the Zeppelin Company itself—changed in fundamental ways. In 1932 Paul Hindenburg was elected the president of Germany, having defeated the runner-up, Adolf Hitler, by a sizable margin. In January 1933, however, Hindenburg appointed Hitler chancellor of Germany. And in 1934, when Hindenburg died, Adolf Hitler became both the chancellor and president of the country.

Hitler, it turned out, was not a big fan of the zeppelin, having once said: "The whole thing always seems to me like an inventor who claims to have discovered a cheap new kind of floor covering which looks marvelous, shines forever, and never wears out. But he adds that there is one disadvantage. It must not be walked on with nailed shoes [in fact, *Hindenburg* mechanics wore rubber-soled sneakers] and nothing hard must ever be dropped on it because, unfortunately, it's made of high explosive. . . . No, I shall never get into an airship."

Hermann Goering, the Reich air minster, also had an aversion to airships. He had been an ace fighter pilot during World War I and was accustomed to aircraft that were small, fast, and highly maneuverable, attributes that did not even remotely pertain to the ponderous and galumphing zeppelins. Both Hitler and Goering, however, as well as propaganda minister Joseph Goebbels, knew that the *Graf Zeppelin* and the *Hindenburg,* on account of their *Kolossal* size, emotional impact, and otherworldly aspect, could function as superb propaganda vehicles for the Nazi

Party. With this in mind, Goering offered Eckener nine million marks toward the construction of the *Hindenburg,* while Goebbels offered an additional two million from the Propaganda Ministry. For that reason, the *Hindenburg* acquired a reputation as "a Nazi airship."

"The airship does not have the exclusive purpose of flying across the Atlantic," Goering had said in a 1935 speech, "but also has a responsibility to act as the nation's representative. . . . I hope that the new ship will also fulfill its duty in furthering the cause of Germany."

When LZ 129 emerged from its Friedrichshafen building shed for its first trial flight on March 4, 1936, it did not yet have a name painted on its hull (another bad luck sign according to shipbuilding folklore). In fact, the craft's name, as well as its location on the hull, had been decided as far back as December 1935, but the name did not get painted on—in red Gothic lettering, as opposed to the officially sanctioned black—until sometime between March 18 and 23, 1936. What the ship did have on its maiden flight was something more important to the Reich than a mere name: four enormous black swastikas on its gigantic, 100-foot-long tail fins. It also displayed the five interlocking and multicolored Olympic rings on both sides of the hull, since the ship, as part of its propaganda work, was scheduled to make a flyover appearance at the 1936 Berlin Summer Olympics.

The *Hindenburg's* first flight was under the nominal command of Captain Hugo Eckener, although Captains Ernst Lehmann, Hans von Schiller, and Max Pruss were also on board to assist. And so, in the midafternoon of March 4, 1936, Pruss gave the order "Airship march," and the ground crew pulled the ship from the construction shed. It came into view at a glacial rate, foot by endless foot, and continued to do so for the next several minutes.

Then, at the further command "Up ship!" ground crew members holding on to the grab rails on both sides of the control gondola effectively threw the leviathan into the air. At the altitude of about 300 feet, the mechanics started the engines, and the ship moved forward. On the ground, observers saw what looked like black smoke streaming from the top of the hull. In reality, this was but an accumulation of dust and dirt being blown off the surface.

Harold Dick, an American engineer working at Friedrichshafen for the Goodyear Tire and Rubber Company, was aboard the *Hindenburg* during its launch, stationed in the control car's navigation room. The craft was steady, stable, and quiet, he thought. "The outer cover, on the other hand, was by no means as quiet as it might have been. I thought additional doping might help, but antiflutter wires and girders would have to be added in quite a few panels." On another flight he reported that "a smell of fuel oil was always present in the keel although there was no sign of leakage."

On March 7, 1936, three days after the LZ 129's first ascent, Adolf Hitler ordered the occupation of the Rhineland.

The Rhineland was an area of western Germany that the Treaty of Versailles had taken away from Germany in 1919, and which the 1925 Locarno Treaty had turned into a demilitarized zone. When Hitler sent troops into the Rhineland, therefore, he was violating the terms of both treaties, but the move was nevertheless wildly popular within Germany. Propaganda minister Goebbels wanted to show the world just how popular the invasion was, and for this purpose he called for a national plebiscite to take place on March 29. It would be a referendum in which the German people could formally express their overwhelming and passionate support for the actions of their führer. Both the *Graf* and the *Hindenburg* were pressed into action to get out the vote in a four-day, marathon "Plebiscite Flight." (Many crew members contemptuously referred to these displays as "circus flights," and Hugo Eckener did so in even worse terms.)

"This flight," said British airship historian John Duggan, "is really the story of two airships, the legendary *Graf Zeppelin* and the new technical monster of the skies, the *Hindenburg,* combining together . . . with the overt political purpose of assisting in conferring legitimacy upon what was an illegal government act."

Captain Lehmann got matters off to a grand and gala start during the launch of the first Plebiscite Flight when, on March 26, he was faced with gusty winds and the prospect of a downwind takeoff—that is, with the wind blowing from the rear of the ship. The idea was to have the stern ground crew release their lines first, allowing the tail to rise up clear of the ground, and then for the men at the nose to cast off their lines, after which the ship could fly away safely and be gone.

During this procedure, however, Lehmann, for unknown reasons, ordered that water ballast be released from the bow of the ship, a measure that had the effect of tilting the *Hindenburg* backward and causing the tail fin to strike the ground, crumpling up the lower portion. Some zeppelin historians have speculated that the shock of the impact weakened the ship's framework, planting the seed of its final destruction. In his autobiography, Lehmann himself waved away the whole issue, calling the incident "meaningless in itself," blaming "the unpracticed ground crew," and going so far as to say that the mishap "accidentally presented a proof of the dependability of the new type of airship," since, after all, it had otherwise survived intact. This was a textbook case of turning a blunder into a virtue. Whatever the truth, the damage, which was minimal, was patched up in a matter of hours, and the *Hindenburg* got on with its official business of Nazi Party advertising.

So now the *Hindenburg,* accompanied by the *Graf,* the two enormous German sausages, proceeded to motor hither and yon across Germany, littering the countryside with political leaflets dropped out of portholes in the lower vertical fin, sending swastika flags earthward on tiny parachutes, and blasting propaganda messages, martial music, the national anthem, and the "Horst Wessel Song," over loudspeakers that had been specially installed for the event. Captain Lehmann particularly relished the flight over Hamburg, where "for an hour we cruised over the city, greeted everywhere by deafening *Heils!*"

The Plebiscite Flight finally came to an end on March 29, 1936. Then came the actual voting, which for the LZ 129's crew members took place aboard the *Hindenburg* itself, in a polling booth that had been set up on the port promenade deck. "In half an hour," Lehmann said, "the LZ 129 voted; 104 eligible voters on board, 104 votes for *Der Führer.*"

The tally across the country was similar. The ballots had spaces only for "Yes" votes.

A little more than a year later, on the evening of May 3, 1937, the *Hindenburg* embarked on its final flight. Earlier that day, the ground crew had withdrawn the craft from the hangar at Frankfurt. The passengers had arrived from their hotel in the city. Margaret Mather, one of them, was suddenly seized by "a strange reluctance" and by the fleeting thought that passed through her head: *What a beautiful farewell to earth.*

Despite those weird premonitions, she nevertheless boarded the ship, along with thirty-five other passengers who climbed the steps up the narrow, drop-down stairways on either side of the keel corridor. Pursers stowed baggage in the cabins of the passengers while they themselves gathered at the windows in the lounges to watch the casting-off process. Theodor Ritter, an engine mechanic, caught a brief view of the airport cafeteria and for a moment imagined a glass of Dortmund, the famous German beer, sitting in front of him.

After the band played some marches, military tunes, and a hymn, the handlers let go of their hold upon the craft, which thereupon rose into the overcast skies of Frankfurt's Rhein-Main Airport.

There is a small literary genre devoted to expressing what the initial ascent of a zeppelin was like. It was like "falling upwards into the sky," like a curtain rising, like going up in an elevator, like experiencing "an indescribable feeling of lightness and buoyancy—a lift and pull upward, quite unlike the takeoff of an airplane."

The craft glided up in silence to an altitude of about 200 feet, at which point the mechanics stationed in the four separate engine cars started up their loud machines. The combined power of the four immense Daimler-Benz diesels would make for a cruising speed of seventy-five to eighty miles per hour.

Once established on course, the *Hindenburg* flew northwest toward Cologne, guided by beacons that flashed from one hilltop to the next. The passengers watched from the promenade deck windows as the ship sailed over hamlets and villages whose lights gleamed in the darkness. Then, just a little more than an hour after takeoff, they were over Cologne, whose cathedral was bathed in a yellow light. The *Hindenburg* made a mail drop there, releasing sacks of letters and postcards on parachutes. Then it made for the German-Dutch border.

At about 10:00 p.m., the stewards, of whom there were seven aboard, plus a stewardess—one attendant for every five passengers—served a supper of cold meats and salads in the dining room on the upper deck. This was a long, red-carpeted area with tables covered in white linens, each flanked by light-weight aluminum chairs designed by the architect Fritz August Breuhaus, the ship's decorator.

After the meal, some passengers retired to their cabins for the night while others collected in either the starboard side lounge on A deck or the bar or smoking room, both of which were on B deck immediately below. During the *Hindenburg*'s first year of operation, the focal point of the lounge had been the aluminum baby grand piano that the Julius Blüthner piano firm of Leipzig specially built for the craft. Although it weighed only 356 pounds and was covered in pale pigskin, it had been removed for the 1937 season.

Later that night, the ship headed out over the North Sea, flew across the English Channel, and finally set off across the North Atlantic.

The chief steward on the trip was Heinrich Kubis, who had joined the Zeppelin Company and started serving passengers in March 1912, aboard the airship *Schwaben*. He had previously worked at a few of the upper-class hotels of Europe, including the Carlton in London and the Ritz in Paris. Early on in the flight, Kubis noticed that the Doehner boys, Werner and Walter, were playing with a toy tank that gave off sparks when they ran it across the floor of the lounge. Kubis explained to them that he had to confiscate the toy for the remainder of the voyage, as sparks were dangerous aboard an airship filled with hydrogen.

Later, Kubis had to intervene again when Joseph Späh, the acrobat, was making unaccompanied trips to one of the rear freight rooms to care for his trained performing dog, Ulla. Kubis informed Späh that such solo excursions through the ship were strictly forbidden, although this didn't seem to matter to Späh in the least, as he continued to make unescorted visits back to the cargo area.

Otherwise, the *Hindenburg*'s final flight was by all accounts a rather humdrum, run-of-the-mill crossing. Leonhard Adelt, an experienced zeppelin passenger, later called it "the most uneventful journey I ever undertook in an airship." Except for the headwinds, that is, which were quite strong, running up to fifty miles an hour, and more or less directly on the nose of the ship. The wind made whitecaps on the waves below. Captain Pruss, seated next to Margaret Mather at the captain's table, told her that, on account of the heavy weather, this was one of the worst trips he'd ever made. He was in fact worried about arriving late in New Jersey, because the *Hindenburg* was scheduled to make a return flight within hours of its arrival and that trip was fully booked with people traveling to England for the coronation of King George VI on May 12.

By the second full day aloft, May 5, the sky had cleared and icebergs, shining white against the cold, gray sea, appeared off the coast of Newfoundland. The ocean water was so transparent that their submerged portions were visible below the surface, in pale green hues. In the sky several vivid rainbows arced toward the sea.

Joseph Späh shot some black-and-white footage of the icebergs with his home movie camera. He also captured several of his fellow passengers on film, including Peter Belin, Claus Hinkelbein (a Luftwaffe officer), and Captain Ernst Lehmann. At some point Späh must have handed the camera to one of the others, since he himself appears in the film too, leaning so far out of an open window that he almost loses his hat.

Finally it was May 6, the last day of the voyage. For the American passengers, this was the most exciting part of the trip, as they could now recognize familiar landmarks and the other sights of their homeland and point them out to others. Peter Belin identified Yale, his alma mater. Emma Pannes spotted the bay on Long Island where she lived with her husband John.

The last of the planned, formal meals had been served the night before, prepared by head chef Xaver Maier and his staff of five. It had consisted of cream soup, stuffed veal cutlet with Swiss asparagus and smoked Westphalian ham, and cucumber salad in cream sauce, followed by a mixed compote. Now, owing to the ship's late arrival, the chef served an impromptu lunch, which was followed later on by sandwiches.

At about three o'clock, the city of New York came into view and the passengers marveled at the tall buildings. Leonhard Adelt later remembered that "in the mist the skyscrapers below us appeared like a board full of nails." The boats in the river sounded their horns, which could barely be heard. The Statue of Liberty seemed like a small, white porcelain figure.

And then they approached Lakehurst, their landing site. There were ominous dark clouds all around and puddles on the field, plus driving rain and occasional flashes of lightning.

"Not at all dangerous," someone said. "A zeppelin can cruise about indefinitely above the storms. It's not like a plane, which has to come down for fuel."

Somebody else told the story about the *Graf Zeppelin,* which had once arrived at its port of call in South America, where a revolution was in progress. The craft had simply milled around off the coast until the fighting was over, after which it landed normally.

At Lakehurst, the landing crew was nowhere to be seen. Cars were parked around the airfield's perimeter, however, and several people on the ground were waving to the ship.

Because of the headwinds en route, the *Hindenburg* was now fully twelve hours late, and Captain Pruss wanted to land as soon as it was safe to do so—as did many of the passengers, since Lakehurst was nobody's final destination. Everyone had connections to make or drivers, friends, or relatives standing by at the field to pick them up.

Nevertheless, it was not best practice to land a hydrogen zeppelin with a thunderstorm in progress, and so it now became the *Hindenburg's* turn to mill around off-coast until the weather conditions improved. Pruss set a course for the town of Tom's River, and then to Asbury Park on the shore.

On the port side of the craft, which is the side of the *Hindenburg* that faces the viewer in the canonical newsreel footage of the disaster, Joseph Späh, Peter Belin, Margaret Mather, and Marie Kleeman, among others, were lined up at the observation windows. The Doehner family—Hermann, Matilde, and the three children—were on the same side farther back, more or less together in a clump. On the starboard side, at the front, were Nelson Morris and his pal Bert Dolan and, at the rear, Gertrud and Leonhard Adelt.

The passengers watched the scenery slide by below them as the craft sailed in a leisurely fashion over the mostly deserted New Jersey beaches. At about this time, stewards came around with tea and sandwiches, the very last fare to be offered aboard the *Hindenburg*.

At the US Naval Air Station in Lakehurst the person in charge of landing operations was Commander Charles E. Rosendahl. This was the same Charles Rosendahl who, in September 1925, while trapped inside the forward section of the broken-up naval airship *Shenandoah,* had been lifted to the altitude of 10,000 feet before descending back again to earth, onto which he crawled out all in one piece. Despite that harrowing experience, in which fourteen of his fellow airmen had died, Rosendahl had always remained a staunch fan and supporter of the airship, and always would.

At 6:12 p.m. local time, Rosendahl sent a somewhat puzzling radio message to Captain Pruss aboard the *Hindenburg*:

CONDITIONS NOW CONSIDERED SUITABLE FOR LANDING GROUND CREW IS READY PERIOD THUNDERSTORM OVER STATION CEILING 2000 VISIBILITY FIVE MILES TO WESTWARD SURFACE TEMPERATURE 60 SURFACE WIND WEST-SOUTH-WEST EIGHT KNOTS GUSTING TO 20 KNOTS SURFACE PRESSURE 29.68.

At first glance it was difficult to reconcile "suitable for landing" with "thunderstorm over station," but the *Hindenburg* was at that point in the vicinity of New Gretna, New Jersey, some thirty miles south of the airfield, and what Rosendahl meant, and which Pruss understood, was that by the time the ship arrived back at the field the storm would have moved a safe distance away. Pruss therefore turned the ship around and headed for Lakehurst.

As they floated over the Pine Barrens of southern New Jersey those at the observation windows could see deer darting around through the sparse vegetation below. In the officers' mess, Werner Franz, the cabin boy, was washing some plates and putting them up on shelves in a cupboard. When he looked out one of the windows, he saw on the ground a couple of kids on bicycles racing to keep up with the zeppelin.

Then, at 7:08 p.m., Rosendahl radioed:

CONDITIONS DEFINITELY IMPROVED RECOMMEND EARLIEST POSSIBLE LANDING

Soon Lakehurst was again visible through the mist, and at about this same time the air station personnel caught sight of the ship.

And then, at last, the *Hindenburg* was over its final destination. The passengers could now see the hangar clearly, as well as the mooring masts and the ground crew, a total of 230 people: 92 navy men and 138 civilians.

On its initial approach, from the southwest, the *Hindenburg* was at a height of 600 feet. To lose altitude, Pruss circled the field to the left and valved hydrogen for fifteen seconds from several of the cells.

There were two mooring masts at Lakehurst, and the *Hindenburg* headed for the one nearest the hangar. The ground crew had collected around that mast. At that time the wind was from the east, and so Pruss approached from the west, so as to be heading into the wind for the landing, as was standard operating procedure.

But Hugo Eckener, in his 1919 "Practical Hints for Piloting Zeppelin Airships," had written that "every landing turns out differently, is an entity in itself, which the commander must fashion according to circumstances," and Pruss did so now, for after he had circled the field and set up an eastbound approach to the mast, the wind changed direction and swung around to the south—in other words, toward the craft's starboard side. This meant that the ship had to change its course in response and make its approach from the north. So Pruss now ordered an S-turn, first swinging left, out and away from the mast, so as to put himself to the north of it, and then turning sharply right in order to face the mooring mast head-on and be lined up directly into the south wind.

On the ground, cameramen from five different newsreel companies—Fox Movietone News, Hearst, Pathé, Paramount, and Universal—were on hand to record these events, in addition to a handful of still photographers, many of them standing on the roofs of cars. In a display of exceptional misjudgment, the

Universal cameraman left the site on account of the bad weather
and took himself into Manhattan to attend a Broadway play. The
others remained and arranged themselves near the mooring tower,
close to the ground crew and the customs inspectors.

Separately, an audio description of the landing was being nar-
rated by Herbert Morrison, for broadcast the next day on the Chicago radio station WLS, "the Prairie Farmer Station," where he
worked as an announcer. His words were being recorded on a large
revolving acetate disk by WLS sound engineer Charley Nehlsen.

Because of the rain, Morrison was standing inside the hangar.
He spoke in a suave, polished voice. "We are now broadcasting
this description, just around sunset, the end of twilight right now.
And raining, raining, as hard as could be. . . . We've been told that
the airship is going to make an attempt at landing in the rain, and
if that's the case we're going to have a mighty fine description of
it for you."

As the ship approached more closely Morrison left the hangar
and ventured out into the drizzle. "Well, here it comes, ladies and
gentlemen, we're outside now, outside of the hangar. And what a
great sight it is, a thrilling one, a marvelous sight."

In the control car, Captain Max Pruss had divided up responsi-
bilities for the approach and landing phases of the flight between
himself and his first officer, Captain Albert Sammt. Pruss would
be responsible for speed and heading, while Sammt would be in
charge of altitude and trim, which were controlled by elevator and
by the selective, alternating releases of hydrogen gas and water ballast. Just before the S-turn, Sammt had noticed that the ship was
tail-heavy and out of trim by several degrees, and so he ordered
that hydrogen be valved from six of the front gas cells in order to
remedy the imbalance and bring it back to level. After two such releases, however, the ship was still tail-heavy. Sammt now dropped

ballast from the rear of the ship to lighten the aft section—three consecutive dumps of water from the ballast tanks at ring 77.

Sammt leaned out of a control car window and watched the water fall to the ground in consecutive feathery streams. There had been thunderstorms in the vicinity, he knew, and he wondered briefly whether it was possible that the cascade of falling water might ground the ship electrically, causing a static spark or some sort of electrical disturbance. Then he dismissed the thought.

"The ship is no doubt bustling with activity," Morrison said. "Orders are shouted to the crew, passengers lining the windows looking down at the field ahead of them, getting a glimpse of the mooring mast."

On the newsreel film, one can see the *Hindenburg*'s three successive water drops and observe the ship drifting laterally to the east as it completes its sharp right turn toward the south, to line it up with the mooring mast and the wind. But even after those releases of water, which totaled more than 2,400 pounds, the ship was *still* heavy at the stern . . . as if, just possibly, hydrogen had been leaking from the rear cells and was continuing to do so. Sammt therefore resorted to *human* ballast, ordering that six crew members move from the center of the ship to the bow: "*Sechs Männer nach vorne!*" he shouted into the telephone ("Six men forward!").

"It's coming down out of the sky, it's pointed directly at us," Morrison said.

Then, for a moment, the *Hindenburg* appeared to be motionless, hovering in place over the ground.

"It's practically standing still now," Morrison said. "They've dropped ropes out of the nose of the ship, and they've been taken a hold of down on the field by a number of men."

Then, in quick succession, several things happened.

One of the ground crew members, a navy man named R. H. Ward, saw that the fabric high up at the rear of the ship was fluttering, as if gas was escaping from the hull. Farther away, and in fact standing outside the air station boundary, Mark Heald, a Princeton history professor, was watching the landing together with his wife and son. Suddenly he saw what appeared to be St. Elmo's Fire, a luminous electrical discharge, running down the spine of the ship just forward of the vertical fin at the top.

"Oh heavens, the thing's afire!" he said. "Up along the top ridge!"

On the ground, Willi von Meister, the Zeppelin Company's American representative, was looking at the aft end of the ship when he saw the interior of the gas cell just forward of the rear fins light up brightly, "like a Chinese lantern," he said later, and then burst violently into flame. There was a loud explosion, like a clap of thunder.

"It burst into flames!" Herbert Morrison cried out. "It's fire— and it's crashing! It's crashing terrible!"

The whole aft end of the ship was suddenly engulfed in flames. The stern dropped abruptly while the nose pointed up. And then the whole burning structure descended, like an airborne bonfire falling through space.

The ground crew scattered amid screams and shouts.

"This is terrible," Herbert Morrison cried out. "This is one of the worst catastrophes in the world."

Chapter 8

FROM HUBRIS TO HORROR IN THIRTY-FOUR SECONDS

Aboard the *Hindenburg,* the first person to note that anything was amiss seems to have been Helmut Lau, one of the helmsmen. Crew members were assigned specific landing stations, points on the ship where they had to be located prior to mooring. Lau had been in the crew's mess room on B deck eating dinner when he heard the landing signal: one long bell, two short bells, followed by one long bell. He soon traveled back to the rear of the craft and descended a ladder leading to the auxiliary control room on the starboard side of the lower fin, where he was responsible for taking variometer readings. Lau was looking up toward rigger Hans Freund, who was above him on the keel walkway, when he heard "a muffled detonation" and saw a bright reflection on the front of gas cell 4, about fifty or sixty feet from where he was standing. Inside the cell he saw red and yellow flames and smoke—and then cell 4 simply vanished before his eyes, destroyed by the heat.

Hans Freund, who was in the process of paying out the stern landing lines, heard "a *whoom*" and was almost immediately surrounded by fire.

Rudolf Sauter, the chief engineer, was also in the lower fin, waiting to receive orders from the control car, when he saw a flash of bright white light at the front of gas cell 4, at about the level of the axial corridor, which is to say, at the centerline of the ship.

Behind Sauter and a little below him, Richard Kollmer was lowering the rear landing wheel when he saw a reddish-pink flash of light above him, which he immediately recognized as fire. Simultaneously he heard a sound that he later described as like "the firing of a gun, a small gun or rifle."

These four men, Lau, Freund, Sauter, and Kollmer, would turn out to be the closest to what has come to be regarded as the starting point or epicenter of the blaze that put an end to the *Hindenburg*. Shortly after the detonation, the men felt a distinct downward acceleration, as if the tail of the ship had fallen out from under them. In fact, the entire aft end of the zeppelin, from frame 140 (the site of the forward engine cars) rearward to the tail cone, had erupted into flames, which now rose into the sky in the form of a single huge fireball. The shock of the initial blast was enough to dislodge two ballast tanks from their mounts. They burst through the floor of the craft and fell to the ground.

"At that same time parts of girders, molten aluminum and fabric parts started to tumble down," Helmut Lau testified later. "The whole thing only lasted a fraction of a second."

Within that second, the explosion's shock wave jarred loose some bits of ceiling plaster in the room where Charley Nehlsen, the audio technician, was recording Herbert Morrison's commentary. It also momentarily upset the cutting head of Nehlsen's "Presto" portable recorder, producing a shallow groove in the acetate disk at the time of the detonation, just as Morrison shouted, "It burst into flames!"

Aboard the craft, many others also heard the sound of the initial explosion, which they later described in a variety of terms and images and reported as having different degrees of loudness, depending on where they were situated within the vessel. To one it was about as loud as a bottle of beer being opened; to another it was like paper bag being popped; to a third it was like the report of a service rifle fifteen to twenty feet away; another said it was "like the noise you hear when you turn a kitchen gas flame on or off."

Farther forward, away from the rear fin by a distance of some 500 feet, the officers in the control car felt rather than heard the detonation. The control gondola, at the time, was full almost to the point of bursting with captains of the line. Six zeppelin skippers were present in the car: Max Pruss, Albert Sammt, Heinrich Bauer, Walter Zeigler, Anton Wittemann, and Ernst Lehmann.

Sammt felt the ship lurch, as if one of the mooring lines had snapped, but when he looked out he saw that they were intact. Then he beheld a chilling sight: bright red flames reflecting back from the windows of the zeppelin hangar. At about that point, Willy Speck, the radioman, shouted "Fire!"

And then the *Hindenburg* simply cracked in half, as if its back had been broken. While the aft portion remained level, the forward section of the ship tilted up sharply at a forty-five-degree angle, while the whole connected flaming ensemble slowly sank toward the ground.

Many of the passengers, meanwhile, were at their own set of windows on either side of the upper deck, where they were watching the landing procedure. Among those on the port side were, front to back, Joseph Späh, Peter Belin, Margaret Mather, Marie Kleeman, and the Doehner family: Herman and Matilde and their three children, Walter, Werner, and Irene. Heinrich Kubis, the purser, was nearby in the port-side dining room, setting up a table for the incoming customs officials.

Peter Belin had been taking pictures of the naval base as the ship made its approach. His roll of film survived the disaster and contained some of the very last images to be taken from aboard the *Hindenburg*. They showed the hangar, indistinct behind a veil of fog and cloud in the dim distance, and then, from directly over-head, a group of ground crew members standing around the land-ing flag.

Belin and Margaret Mather, who were side by side, watched as the landing lines came down out of the nose. Farther back, Marie Kleeman sat calmly and observed events at another win-dow, while the Doehners at the rear of the promenade remained together at first, but then Herman Doehner, realizing that he was out of movie film, left his family for a moment to go to his cabin on the lower deck for a fresh supply.

And then, out of nowhere, the passengers heard the sound of an explosion. Instantly, the floor dropped away and slanted back. This was an almost unbelievable event: given the size and, hitherto, the stability of the ship, it must have felt as if the very world itself had fallen away from beneath their feet. Some of the passengers—Peter Belin and Joseph Späh—held on to whatever was near them to keep from sliding down the steeply angled floor. Margaret Mather, by contrast, "was hurled a distance of fifteen or twenty feet against an end wall," as she later reported. "I was pinned against a projecting bench by several Germans who were thrown after me. I couldn't breathe, and thought I should die, suffocated. . . . Then the flames blew in, long tongues of flame, bright red and very beautiful."

Matilde Doehner and the three children were now at the very bottom of the heap, all of them pressed against the rear wall.

Below them, in the officers' mess, Werner Franz was cleaning up the last of the dishes. He had a coffee cup in his hand and was just about to put it away when he heard a dull thud. Then he felt the ship heave up and slant back, spilling the dishes from the cupboard.

In the control car, Sammt, Pruss, Lehmann, and the others also grabbed on to posts, girders, railings, or whatever else, and held on to keep from falling. Then, as the ground approached, Lehmann yelled, "Everybody to the windows!" The officers braced themselves as the ship struck the earth and rebounded up slightly when the pneumatic landing wheel hit the ground and pushed back. As the ship came down for the second and last time, Lehmann, who was already half out of a window, yelled, "Everybody out!" Finally, at a height of about fifteen feet above ground level, he, Sammt, and Pruss leapt from the gondola. When they hit the sand they ran to starboard, but as they did so the ship itself rolled lightly to starboard and collapsed on top of them.

On the port-side upper deck, Heinrich Kubis yelled for the passengers to jump. But leaping from a burning zeppelin required the most exquisite sense of timing: jump too soon and you might hurt or even kill yourself in the process, but jump too late and you'd probably be burned to death. Worst of all was the fact that you had literally only a matter of seconds in which to size up the situation, make your decision, and act.

Kubis himself sat on a windowsill, waited until the ship approached the ground, and then leapt out, followed by several passengers. Joseph Späh smashed through a window with his movie camera and dropped through the air. Peter Belin broke through another window and followed suit. Margaret Mather "saw a number of men leap through the windows, but I just sat where I had fallen. . . . I was thinking it was like a scene from a medieval version of hell." As that thought passed through her head, she heard a man exclaim, *"Es ist das Ende,"* which was what she herself thought.

Marie Kleeman, in the midst of this chaos—"Everything was noises and shrieks and screams"—nevertheless remained quietly seated throughout, clutching in her hand a pair of gloves.

On the deck below, a water ballast tank broke open and drenched Werner Franz with its contents, making him momentarily immune to fire and flame. But the blaze was coming up toward him through a passageway, and he ran forward to a provisioning hatch and hurled himself through it.

That left the Doehner family, who had been jammed against the rear wall. The ship's forward section leveled out and began to settle gently to the earth, and once it was completely on the ground and stopped, Mrs. Doehner could see someone outside motioning to her and calling. She picked up her little son Walter and dropped him to the man below, who caught the child, and then did the same with Werner. Her daughter, Irene, age fourteen, had gone below, to B deck, to find her father. Finally, Matilde Doehner had no choice, and she herself jumped from the ship.

Simultaneously with those on the port side, the passengers on the starboard side were going through the same sequence of events: they watched the landing lines drop, they heard the dull thud, they felt the floor slant back, they heard the shouts and screams. At the front windows of the starboard promenade stood Nelson Morris together with his former colleague, Bert Dolan, whom Morris had persuaded to board the *Hindenburg* with him so that Dolan could make it home in time for Mother's Day. They saw flashes of lightning off to the west.

A succession of other passengers lined the windows farther back, while at the very rear were the two journalists, the husband-and-wife team of Leonhard and Gertrud Adelt. Leonhard had collaborated with Ernst Lehmann on the book *Zeppelin,* which was in effect the captain's autobiography. It would be published later that year, including a final chapter, "The Last Flight," written by Lakehurst commander Charles E. Rosendahl. Aboard the ship, Leonhard Adelt at first heard "a light, dull detonation." Then, as he remembered later, "I turned my gaze toward the bow and noticed a delicate rose glow, as though the sun were about to rise. I understood immediately that the airship was aflame."

As fire and smoke began to enter the starboard-side dining room, the passengers took their chances and leapt, only to be confronted with the further threat that when the *Hindenburg* struck the earth for the last time it rolled to starboard. Once they landed on the ground, therefore, the starboard passengers had to push their way through the tangle of red-hot girders and wires with nothing but their bare hands.

In the bow of the ship, finally, were twelve men whose location was the worst of all, as it placed them at the very mouth of hell—the modern version. Even when it flew straight and level, the floorboards of the *Hindenburg's* nose section canted progressively ever more sharply upward until they came to a point at the nose cone. The angle of the keel walkway became so steep near the end that its planks had been supplanted, first by a duralumin stairway, and ultimately by ladders that rose to platforms around the nose-cone connecting device, the metal pendant that when fitted into the matching cup-shaped "flower pot" at the top of the mooring mast held the ship securely attached.

The process of coupling the *Hindenburg* to the mooring tower was a tricky operation that normally required a team of six men.

On this occasion, however, the usual six were augmented by the six crew members Captain Albert Sammt had sent forward in an effort to level the ship. When the rear of the craft burst into flames, these twelve men, who included a twenty-six-year-old rigger by the name of Erich Spehl, were arrayed out at various benches, platforms, windows, and hatches lining the zeppelin's foremost tip. With the forward part of the ship angled precipitously upward, the men in the bow faced the choice of being burned alive in a rising ball of flame or jumping out into thin air. The nose of the airship was still far off the ground, and those who leapt risked serious injury or death.

Now, as the rear section crashed to the earth, a large jet of fire streamed upward from the axial corridor, which acted as a flue or chimney, focusing and concentrating the flames and turning the forward section into an enormous midair blowtorch. Standing on the mooring shelf at the farthest extremity of the axial walkway were three men: Alfred Bernhardt, Ludwig Felber, and Erich Spehl, all of whom were immediately engulfed by the overwhelming plume of fire.

On the ground, meanwhile, many of the landing crew members were scattering from the falling, burning, crackling monster that was raining down in pieces upon them. The navy ground crew men, however, were under the command of Chief Petty Officer Frederick J. "Bull" Tobin, who, as it happened, had himself survived a prior airship wreck, that of the *Shenandoah,* the American airship that had broken apart in a storm over Ava, Ohio, in September 1925. Now, twelve years later, he called out to the sailors, "Navy men, stand fast!" Tobin had a loud, booming voice of the kind that demanded obedience, and so, as the ship crumpled to the ground, instead of running away, these men ran toward it, to rescue whomever they could.

All of these events took place—everything from Helmut Lau's first seeing the fire's reflection on gas cell 4, to the *Hindenburg's* forward section tilting back while people held on or slid to the rear, to the ship's striking the ground, rebounding, coming back down again, and then collapsing in upon itself while people jumped out and dropped through the air—all of it happened in the space of thirty-four seconds of sheer horror and fright.

Watching these events unfold in front of his eyes, and helpless to do anything about them, announcer Herbert Morrison, clearly distraught and barely able to speak, wailed into his microphone, "Oh, the humanity!"

I n the newsreel clips of the disaster, one can see tiny stick fig-ures running away from the front of the ship as it neared and then settled to the ground. For the most part, these were not ground crew members, but rather passengers and zeppelin crew members who had jumped from or fallen out of the craft.

In the tail section, Helmut Lau, Hans Freund, Rudolf Sauter, and Richard Kollmer found themselves trapped inside the lower fin, which had collapsed and was now lying horizontally on its left side. But Kollmer noticed that the starboard-side entrance hatch was passable, and so he climbed out, followed by the other three. The four men ran to safety nearly unhurt, Kollmer running so fast that, as he said later, "I could have won an Olympic gold medal."

Most of the port-side passengers, likewise, managed to escape more or less intact. When Joseph Späh leapt, he curled himself up to do an acrobatic safety roll upon landing, but he broke his ankle nevertheless. A navy man grabbed onto him and helped him walk away from the wreck. Späh ended up at the air station's dispensary, or infirmary, a room that would soon be filled with survivors—as well as with others who were even then near death.

Peter Belin landed in the wet sand all in one piece. He looked back up and saw looming over him the whole massive airship engulfed in flames. Other passengers were staggering around erratically, and Belin guided some of them into trucks that were headed to the hangar. A separate space inside it had been converted into a press room, and a blackboard listed the names of survivors as they became known. Belin's parents had come to Lakehurst to meet their son, and they scanned the blackboard again and again for his name, but it never appeared.

Margaret Mather waited for the ship to hit the ground, but it alighted so softly that she missed the event altogether. "Suddenly I heard a loud cry! 'Come out, lady!' I looked, and we were on the ground. Two or three men were peering in, beckoning and calling to us. I got up incredulous and instinctively groped with my feet for my handbag, which had been jerked from me when I fell. 'Aren't you coming?' called the man, and I rushed out over little low parts of the framework which were burning on the ground."

She too wound up in the infirmary, next to a horribly injured man who, because all the chairs were taken, was sitting on a table. This proved to be Ernst Lehmann. "He was seated on a table near me," Margaret Mather said. "Most of his clothes and his hair had been burned off." He was holding a bottle of picric acid and applying it to a piece of gauze, with which he swabbed his wounds. "Not a groan escaped him as he sat there, wetting his burns. It was a strange, quiet interlude, almost as though we were having tea together."

Lehmann would have three separate reactions to the destruction of the *Hindenburg*. As he was led away from the wreckage he kept repeating, almost robotically, the phrase "*Das verstehe ich nicht . . . Das verstehe ich nicht . . .*" ("I don't understand it . . . I don't understand it . . ."). Later, when Leonhard Adelt encountered Lehmann in the infirmary, Adelt asked him what had happened. Lehmann's one-word answer was *"Blitzschlag"* (lightning). The next day Charles Rosendahl visited Lehmann in the hospital. Moments before he died, Lehmann blurted out to Rosendahl: "It must have been an infernal machine." In other words, sabotage.

Like Lehmann, Captain Max Pruss had also jumped from the control car to starboard. Despite being badly burned, he at first tried to help others get away from the fire, but was soon enough himself led away from the burning wreckage. His condition was so critical when he arrived in a hospital that a Catholic priest gave him the last rites.

Sammt was luckier, and although his face, hands, and arms had been burned, he was saved from an even worse fate by an incoming blast of cool air that was feeding the flames. He walked away from the wreckage still in full uniform.

Marie Kleeman never once budged from her seat. It was as if a disaster movie was playing out all around her and she was content merely to take it all in as it happened. When the ship came to rest, one of the rescuers led her down to the lower deck and then out through the same stairway through which she'd originally boarded the *Hindenburg* back in Frankfurt. As a sailor led her away from the crash site, still holding a pair of gloves, she was unhurt other than for some singed hair, a cut lip, and a bruise on her face.

Werner Franz had jumped before the ship rebounded, run to safety as fast as he could, and emerged "wet but alive," as he later said. Heinrich Kubis, similarly, escaped practically without a scratch.

Then there was the Doehner family. Herman, the father, died in the flames somewhere on B deck. The two young boys, Walter and Werner, who their mother dropped to a rescuer, suffered burns and other injuries. They were still screaming with pain and terror as they were brought into the infirmary. As for Matilde herself, she fractured her pelvis when she hit the ground. As she was carried away by two crew members, one of whom was Heinrich Kubis, she called out the name of her daughter, Irene.

Matilde had last seen Irene as she was headed toward B deck looking for her father, Hermann. Irene would have died in the wreck, as her father did, had it not been for the heroism of a bystander, Emil Hoff. Hoff was a Veedol Oil Company tanker truck driver who had been waiting at Lakehurst to refuel the *Hindenburg* for its return flight to Germany. Wearing a white uniform, hat, and tie, he had already entered the burning wreckage once in search of survivors. He had encountered the electrician Philipp Lenz, whom he'd led out to safety. Then *he returned to the ship*, entering through a B-deck window, and climbed the stairs to the port-side dining room. As reported by Patrick Russell on his erudite blog Faces of the Hindenburg, "There he found Irene Doehner sitting in a daze at one of the tables. She was badly burned and in shock and Hoff evidently decided that lowering her down the gangway stairs and out through the bottom of the ship (as was being done with other passenger survivors) would take too long and might risk injuring her further. So he led her to one of the dining room windows and tried to get her to jump. The window was still about 15 feet above the ground, and rescuers were still on the ground below."

Irene Doehner, hair and clothes ablaze, stood framed by the dining room window as if in the final scene of a mad, demonic opera, only this was real life.

Finally she jumped to the ground, where several others extinguished her flaming clothes and hair. Still, when she arrived at

Point Pleasant Hospital she was so badly burned that one of the nurses fainted upon seeing her. That night Irene Doehner died.

Of the starboard-side passengers, Nelson Morris had to push through a maze of burning girders, wires, and ring sections with his bare hands—the burning structures separated and broke away "like paper," he later said—but he made it out alive. Burtis Dolan, his friend, did not.

Leonhard and Gertrud Adelt, the journalists, jumped from a height of about fifteen feet, hit the ground in a condition that left them able to run, which they did—"like an automaton," as Gertrud later said. They both escaped with minor injuries.

Farthest forward, in the nose section of the *Hindenburg,* and highest off the ground, were the three men on the mooring shelf at the very tip of the axial walkway: Alfred Bernhardt, Ludwig Felber, and Erich Spehl. It was they who were standing directly in the path of the blowtorch formed by the axial corridor when the hydrogen-air mixture inside it ignited and gushed from the nose as a well-defined cylinder of flame. Given their circumstances, these three men had the least chance of escaping the conflagration alive.

Perhaps made insensible by the flames, all of them rode the craft down to the ground, where rescuers pulled each of them from the burning wreckage. Felber and Bernhardt were so badly burned as to be unrecognizable, and both of them died during the night.

That left Erich Spehl. He too was hideously burned, but when he arrived at the infirmary he was still conscious and able to speak. He asked to send a telegram to his girlfriend back in Germany. Joseph Späh, who spoke both English and German, took down the girlfriend's name and address, as well as Erich Spehl's two-word message, which was: *"Ich lebe"* ("I live"). Späh turned away to send the telegram. And as he did so, Erich Spehl died.

Hugo Eckener was in Graz, Austria, when he first got word that the *Hindenburg* was no more. Eckener was by this time a household name all over Europe, and he was in Graz to give a couple of invited talks, one to the members of a local flying club. He ended his after-dinner remarks with the aviator's traditional parting salute: "Happy landings!" Soon afterward, he returned to his hotel room and retired for the night.

He was awakened about one in the morning by the sound of the telephone. It was the Berlin correspondent for the *New York Times*, a man who identified himself by the name of Weyer. Weyer told Eckener that he had bad news: the *Hindenburg* had exploded and crashed upon landing at Lakehurst. Did Eckener have any statement to make?

Exploded? He could hardly believe this! No, he had no "statement."

Weyer asked if sabotage could have been responsible. With no additional facts at his command, Eckener said he had no idea, but conceded that if the ship had truly "exploded" (which it had not), it was at least a possibility. From that point on, the concept of sabotage became inextricably linked with the *Hindenburg* disaster, although with scant more basis in fact than Eckener had had that night—or than Lehmann had when he spoke to Rosendahl of an "infernal machine."

Eckener's first priority was to get to the accident site, where he could conduct a physical examination and analysis of the ship's remains. Overnight the Reich Air Ministry had put together a German investigating committee and made arrangements for the members to travel to the United States aboard the steamship *Europa*, which had already left from the port of Bremen for New York. The committee included Hugo Eckener, Ludwig Dürr, chief designer of the *Hindenburg*, and Max Dieckmann, a professor at the Technical University of Munich and an expert on static electricity. They and three others were to meet up and board the *Europa* when it stopped at Cherbourg.

Before embarking, Eckener had written a note to his wife saying, in part, "I deeply regret that I was persuaded to allow the use of hydrogen when the *Hindenburg,* after all, had been designed for the use of helium."

Five days after the disaster, on Tuesday, May 11, the bodies of the twenty-eight German dead, including that of Captain Ernst Lehmann, lay in caskets that were draped with swastikas, covered with sprays of flowers, and arranged side by side in a long, straight row on the pier of the Hamburg-American line in New York. During the memorial service, which was attended by some 10,000 members of German organizations, uniformed German soldiers and officers flanking Lehmann's flower-bedecked coffin gave the straight-arm Nazi salute. Afterward, the caskets were loaded aboard the steamship *Hamburg* for their return to Europe.

In the end, of the ninety-seven souls aboard the *Hindenburg,* thirty-five died. Thirteen passengers out of the thirty-six on board perished, as well as twenty-two of the sixty-one crew members. In sum, that amounted to about one-third of the whole in each case. In addition, one member of the ground crew, Alan Hagerman, a line-handler from Lakehurst, also died, bringing to thirty-six the total number of deaths caused by the *Hindenburg* fire. It was remarkable, given the circumstances, that about two-thirds of those aboard survived the crash.

The *Hindenburg* was a ship whose cataclysmic ending brought to final fruition the arguably insane nature of its design, construction, and use as a passenger transportation vehicle. This was a vessel one-sixth of a mile in length, in which seven million cubic feet of explosive hydrogen was used to convey as few as three dozen passengers at a time, putting their lives and safety at continual risk for the full duration of every trip. It had an ending whose ferocity—from the sudden initial flash of brightness in the stern, the fire's subsequent eruption as a mushroom cloud of flame, and its rapid, cell-by-cell spread forward to its final, garish emergence as a whirling column of flame issuing from the axial corridor at the nose—counterbalanced, in its way, the uncanny good luck of its predecessor, the *Graf Zeppelin,* a craft that, though essentially of the same high-risk design and construction, nevertheless never hurt a soul.

The *Hindenburg* disaster was payback for years of consistent bad judgment, as well as for the hubris of imagining that hydrogen airships could be operated safely and with impunity, for an indefinite period of time, routinely, as a matter of course. The disaster was the wages of a pathological technology. But it was far from the only one in history. Others were to come in the years that followed, while yet further examples may lie in our near or distant future.

Most ironic of all was the fact that the *Hindenburg* did not mark the end of the hydrogen zeppelin, not even in Germany. Despite the disaster's extremely palpable nature as an object lesson in human stupidity, in the offing was yet more to come.

PART II

PATHOLOGICAL
TECHNOLOGY'S
CHAMBER OF HORRORS

Chapter 9

PROGRESS THROUGH H-BOMBS

It is in the nature of pathological technologies that they are characterized by grandiose ambitions driven by emotional, romantic, starry-eyed mind-sets or utopian spells. Moreover, proponents of such a technology often wildly exaggerate whatever virtues it actually might possess while at the same time routinely neglecting, disregarding, denying, underestimating, or even ignoring the technology's potential drawbacks, unwelcome consequences, or unintended negative countereffects. The hydrogen airship is arguably one of the most obvious and graphic instances of a pathological technology. Its significance was magnified by the *Hindenburg* fire's having been one of the first great disasters to be captured live, on film, as it happened. But the recent history of science and technology provides several other paradigm cases of the same unfortunate habits of mind at work, a collection of skewed mental states, heightened emotions, and faulty assumptions that have led to palpably inappropriate uses of applied science. One example is the idea of carrying out planetary engineering projects by means of thermonuclear or atomic devices—that is, digging canals, moving mountains, or creating instant harbors by the systematic detonation of one of more nuclear weapons.

That concept, admittedly, embodies a certain amount of intuitive appeal. A bomb, after all, is essentially concentrated energy, and energy is what you need to effect any large-scale physical alteration or displacement of matter. In addition, bombs are quite handy when used for such purposes: they're relatively compact, light, portable, and fast-acting. They are high explosives writ large. Although special problems arise when the bomb in question emits hazardous amounts of radiation and radioactive debris— otherwise known as fallout—worries about fallout never stopped the world's would-be nuclear earthmovers from proposing a succession of mildly deranged building projects, such as blasting a new canal across Central America (the "Pan-Atomic Canal"), excavating a superhighway route across the Andes Mountains and the Amazon Basin, or diverting the course of the Tennessee River so as to connect it up with another one nearby, the Tombigbee, to create a new commercial shipping lane (the so-called "Tenn-Tom Waterway"). All of these projects, and many more besides, were mere grist for the planetary nuclear engineering mill.

Scientists had entertained the notion of using atomic bombs for peaceful purposes even before the detonation of the first such device at Trinity Site in 1945. Later, in 1950, the Los Alamos physicist Frederick Reines (winner of the 1995 Nobel Prize in Physics for his detection of the neutrino) formally assessed the prospect of using nuclear devices to carry out mega-scale civil engineering projects. Writing in the *Bulletin of the Atomic Scientists,* Reines asked the superficially reasonable question: "Are There Peaceful Engineering Uses of Atomic Explosives?"

His answer was, yes and no. "For example, if it is desired to divert a river by blasting a large volume of solid rock, this might be accomplished by detonating an atomic bomb deep in the rock," he said; he also noted, however, that "the local deposition of radioactivity would limit subsequent work in the vicinity." A second shortcoming was that "a hill about one quarter mile high and a quarter mile across could probably be blown apart, but one significantly larger would not be destroyed." (Reines was writing before the advent of the hydrogen bomb, which could obliterate even more sizable mountains.)

The biggest obstacle to more widespread use of atomic bombs in geoengineering, he said, was radiation: "Large quantities of radiation . . . limit the possible uses of atomic explosives as a blasting agent." But then he added the general proviso that, "of course, if the radioactivity is overlooked, more things become possible."

If the radioactivity is overlooked. Yes, that was the key to progress in nuclear geoengineering technology, all right: sweeping under the rug any and all unpleasant consequences.

And so America's cadre of atomic earthmovers tended, by and large, either to overlook radiation or at all events to downplay it, minimize it, underestimate it, deflect attention away from it, or otherwise treat it as a non-issue.

In July 1956, six years after the Reines analysis appeared, Egypt's Gamal Abdel Nasser nationalized the Suez Canal, which was later closed to traffic for several months. At what was to become the Lawrence Livermore National Laboratory (but which was then known as the University of California Radiation Laboratory at Livermore), a physicist by the name of Gerald Johnson toyed with the idea of excavating a new, alternate canal by means of hydrogen bombs, which had recently been demonstrated in the Pacific.

"We looked at the topography of Israel, and what we thought we could do with nuclear explosions, and concluded we could dig a sea-level canal—from a purely technical point of view—all the way across Israel, from the Gulf of Aqaba to the Mediterranean."

That was just an idea, a thought experiment, not a formal plan of action by any means, but it nevertheless marked the beginning of a trend among a small group of scientists who harbored out-sized nuclear geoengineering ambitions. And why not? In addition to the Suez Canal blockage, traffic was increasing in the Panama Canal, which, sooner or later, would itself be ripe for replacement or for supplementation by a new, competing ocean-to-ocean waterway. And so it was no surprise that in February 1957, when the Livermore lab hosted a three-day symposium on the subject of nuclear engineering projects, one of the items under discussion was a scheme for blasting a new sea-level canal across Central America. The Sandia Corporation's physicist Luke J. Vortman, in a paper entitled "Estimated Nuclear Explosive Requirements for Canal Excavation," calculated that 26 bombs with a total yield of 16.7 megatons would suffice to create this new pan-American seaway. Later estimates suggested that as many as 262 to a high of 764 nuclear bombs might in fact be needed for the task, depend-ing on the location of the canal across the isthmus, its length, width, and depth, and the yield of the nuclear devices used. This range of estimates represented a considerable window of uncer-tainty. Nevertheless, in all of these scenarios the radiation hazard, whatever its magnitude, was supposed to be dealt with by the temporary displacement of local residents.

Apparently this was an idea with legs, because in June 1957, four months after the symposium, the Atomic Energy Commis-sion (AEC) established a program for the purpose of systemati-cally exploring the theory and practice of nuclear earthmoving.

These authorities christened the program "Project Plowshare." According to some accounts, the name came from physicist Isidor Isaac Rabi, one of the original atomic scientists, who, when informed of the enterprise by one of the participants remarked: "So, you want to beat your old atomic bombs into plowshares."

That was the idea. The research for Project Plowshare would take place at the Livermore lab, and the principal investigators would be Gerald Johnson, Edward Teller, Ernest Lawrence, and Harold Brown (who was later to be defense secretary under US president Jimmy Carter). Collectively, these men wanted to take atomic or hydrogen bombs, objects almost universally reviled by scientists and laypeople alike, and turn them into something that was recognizably *good.**

To demonstrate the practicality of their scheme the Plowshare scientists needed to undertake one or more acts of benevolent nuclear construction. One of the simplest proofs of concept, they decided, was to blast out a harbor at some remote and uninhabited corner of the Earth. That could be done with a mere handful of bombs—five or six—set in a row; exploding them simultaneously would leave a deep trench in the earth that would automatically backfill with seawater.

Easy enough. But where to perform this amazing feat?

After considering locations as diverse as Arica in Chile, Canada's northern provinces, and Christmas Island in the Pacific,

---

*To describe their bombs, the Plowshare scientists tended to use the catch-all phrase "nuclear devices," which covered equally both hydrogen (fusion) and atomic (fission) bombs. However, as Freeman Dyson has explained, "to minimize the contamination of the landscape by radioactive fallout, the Plowshare experts designed bombs whose explosive yield came mostly from fusion and as little as possible from fission." Thus the strong preference among them for using hydrogen bombs for their nuclear excavation projects.

among other places, the Plowshare team settled on a stretch of apparently barren and deserted shoreline along the northwest coast of Alaska (which at that time was not yet a state). Then they hired the E. J. Longyear Company, a mineral exploration and mining firm in Minneapolis, to narrow down the list of possibilities to one. Longyear settled on a rather craggy and forbidding landmass known as Cape Thompson. Among its virtues, they said, was the fact that the land was already owned by the federal government; in addition, the site had commercial development potential since it was within striking distance of vast coal and oil reserves.

But there were some fuzzy edges to this otherwise bright and rosy picture. In the first place, the area in question was less than forty miles from two inhabited Eskimo villages, one at Point Hope, the other at Kivalina. In the second place, the site was within 200 miles of the Soviet Union, practically on the doorstep of the nation's Cold War enemy. And as for those allegedly bottomless oil and coal deposits, they were in actuality a considerable distance away from the blast site, and no form of transportation between these mother lodes and the seashore, other than dogsleds and teams of huskies, yet existed. Moreover, even if a railway was built to take advantage of this gold mine of opportunities, the coastline in question was ice-blocked for nine months out of the year.

Despite these drawbacks, which were not minor, Cape Thompson would become the setting for what came to be known as Project Chariot.

In the March 1960 issue of *Popular Mechanics,* Dr. Edward Teller (popularly known as "the Father of the H-Bomb") published an article entitled "We're Going to Work Miracles." In it he explained for the masses the general rationale behind Project Chariot. Teller cited the veins of coal "and, somewhat farther, oil that might attract commerce except for one vital lack: There is no good harbor, no good anchorage for seagoing vessels." The obvious solution, he said, was simply to create one, an *artificial* harbor.

"The harbor will be excavated in an instant—in a matter of milliseconds—by the explosion of five nuclear bombs," he said. "In that tiny interval of time this energy will move 20 million tons of earth and rock. It will blast out a channel 1,800 feet long and 750 feet wide and at the same time create an inner harbor a quarter of a mile wide and half a mile long. The minimum water depth will be around 30 feet."

Definitely, this was a new frontier in speed-geoengineering.

Some radiation would be released in the process of creating the "instant harbor," Teller conceded, but most of it would remain in the ground as a result of the bombs having been buried deep below the surface. "We expect that all except 10 or 20 percent of the radioactive byproduct will be trapped at the deep zero points and we hope that it will remain practically immobilized in the fused rock." Well, one can always hope.

This particular miracle would be performed, Teller said, "at the mouth of Ogotoruk Creek." All he had to do was to convince the Eskimos in the vicinity to allow the atomic explosions to take place, since they in fact held proprietary rights to and control over use of the land, even though it was under the formal jurisdiction of the Bureau of Land Management.

Blowing things up would be the easy part; getting the Eskimos on board the project was something else again. For that matter, even persuading the wider general public, particularly native Alaskans, that excavation by means of radiation-spewing atomic or thermonuclear devices was a good idea proved to be a considerable challenge all by itself. To begin with, there were linguistic barriers to be overcome: all of the forbidding, military, nuclear-Armageddon terms such as "warheads," "bombs," "devices," "fallout," "yield," and, especially, "ground zero" made a mere construction site sound like a battle zone.

And so, in 1958, at the very beginning of the program, a team of Livermore public relations professionals got together to concoct a more palatable and inviting nomenclature. Thus, "bombs" became "assemblies," or "nuclear sources." "Fallout" became the more homely "debris." "Yield" was renamed "energy release," while "ground zero" was rechristened "the detonation point."

That same year Teller (who was by then director of the Livermore lab) and Gerald Johnson (he of the alternate Suez Canal concept) flew first to Juneau, and then on to Anchorage and Fairbanks, to drum up support among local prodevelopment groups. The scientists touted the economic benefits to be derived from magically nuking a harbor into being. Doing so would open up the area to oil extraction, mining, and even commercial fishing, which hitherto had been stalled by the lack of a safe haven for fishermen to sail their boats into during bad weather. Besides, the AEC was going to spend $5 million on the project, "two-thirds in labor and things we will buy here," Teller said. How could anybody pass up that kind of windfall? Teller quipped, furthermore, that the nuclear engineers were in such precise control of their devices that they could "dig a harbor in the shape of a polar bear if so desired."

Pithy wisecracks like that were a Teller trademark, but they left a good impression, and many Alaskan businessmen and newspaper editors bought into the scheme. An editorial in the *Fairbanks Daily News-Miner* said: "Alaska has been invited by the high echelon of nuclear scientists of our nation to furnish the site. . . . We say to Dr. Teller and his associates: Come ahead."

Teller did, returning to Alaska many times, once to receive an honorary doctorate from the University of Alaska at Fairbanks and deliver a commencement speech in which he served up yet more of his astounding facts and epigrammatic tidbits. As regards "this hysteria about fallout," he said that a managed dose of it "contributes to radiation less than the wristwatch that I am wearing on my wrist." A month later, during a news conference in Anchorage, a reporter asked him to explain the meaning of his oft-used phrase "geographic engineering." "If your mountain is not in the right place," Teller said, "just drop us a card."

But during all of his visits to Alaska, never once did Edward Teller visit the Eskimo village of Point Hope, which was a scant thirty-one miles from the proposed site of the forthcoming harbor at the mouth of Ogotoruk Creek. Nor did he ever lay eyes on the actual creek itself, except for a brief flight over it in a twin-engine Beechcraft in June 1959. Dan O'Neill, who wrote a book about Project Chariot called *The Firecracker Boys*, said of Teller's fly-over: "Perhaps in his mind he saw a kind of reenactment of the third day of creation: his awesome explosion demarcating what would be land and what would be sea."

In July 1959, the Los Angeles engineering firm Holmes & Narver began work on the Ogotoruk Creek base camp, hauling in building materials, fuel oil, bulldozers, drilling gear, and other supplies on tugboats and barges. The compound that ultimately arose at the site was in effect a mini–Los Alamos, with sleeping accommodations in Jamesway huts (Quonset huts designed for Arctic weather conditions) to house eighty-four scientists and day laborers, plus gasoline generators, water purification equipment, a mess hall, a biology lab, and an airstrip, among other amenities. At the entrance to the complex, workers erected a wooden sign that announced: HOLMES & NARVER. ENGINEERS-CONTRACTORS. PROJECT CHARIOT.

The geologists then proceeded to take a series of core samples— or tried to. Their goal was to drill down through 1,000 feet of earth, but the ground, which was frozen solid, offered so much resistance to penetration that the men quit before reaching 600 feet into the permafrost. After some months of these efforts, the scientists loaded what samples they had, along with their worn-out drilling equipment, onto a barge, which then left for Seattle. But as if this were a Clive Cussler novel, the barge and its tugboat ran into such heavy weather that the barge capsized, sending all the laboriously extracted core samples to a watery grave in the Gulf of Alaska.

Meanwhile, a diverse group of biologists, including some from the University of Alaska and others from the AEC's own Division of Biology and Medicine, plus reinforcements from various other places, arrived at the Ogotoruk Creek site to perform baseline studies of the area's flora and fauna and to assess their vulnerability to whatever amounts of radiation might yet manage to escape from the coming precision bomb-blasts. While Teller and company had initially regarded the area as remote, desolate, and essentially barren, AEC biologist John Wolfe gave a talk ("Ecological Aspects of Project Chariot") at the 1959 Second Plowshare Symposium, in San Francisco, in which he pointed out that the blast site "is not *remote* to the Eskimo, the arctic fox, the ptarmigan," and that "the flowering plants blooming by the thousand per acre in brilliant colors during the early period of wet earth belie its *barren*-ness and *desolation*," and so on.

Finally, in March 1960, the same month in which Teller's *Popular Mechanics* "We're Going to Work Miracles" piece came out, the AEC belatedly took notice of the fact that there were actually *people* living at Point Hope and decided that it might be politic for someone at the commission to pay them an official visit. And so, for the first time, they sent two AEC reps to Point Hope: Russell Ball and Rodney Southwick, along with parasitologist Robert L.

Rausch, who lived in the area and was known to and friendly with the native Inupiaq.

In the afternoon of March 14, 1960, the three men confronted an audience of about 100 Eskimos in one of the village's public buildings. The Eskimos were seated on the floor, as was their custom. Because they had no written language, and since the telephone had not yet reached this remote, barren, and desolate northern outpost, it was a common practice among the Inupiaq to communicate with distant relatives by means of tape recordings. Many in the audience, therefore, were equipped with these machines, which is how it came about that everything said during the meeting was taken down and immortalized on magnetic media, word for word. For example: "Thank you, Mister Frankson. Ladies and gentlemen of Point Hope, we have come here as representatives of the Atomic Energy Commission," and so on.

Their general message was: Not to worry, everything's cool, everything's going to be done safely, there will be no interference with your caribou hunting, fishing, and egg-collecting, which we understand you have been doing from time immemorial in and around the blast area. As for radiation damage, it is *just not going to happen*. "At this distance," Russell Ball said, "the amount of airborne radioactivity which could reach here would be, could not *possibly* be enough to cause injury to people or animals. There's just no chance of that."

But the effect, if any, of all of that mollifying rhetoric was pushed aside in short order by the showing of a film that the AEC reps had brought along with them to Point Hope. Big mistake! The film had been made at Livermore two years earlier, in 1958, for screening at the Second International Conference on Peaceful Uses of Atomic Energy in Geneva. It was called *Industrial Applications of Atomic Explosives,* and it graphically depicted, by animation, how an "instant harbor" could be created by means of shallowly buried nuclear bombs.

Despite its being a simulation, the area was fully recognizable as the mouth of Ogotoruk Creek at Cape Thompson, with the hills, cliffs, and shoreline of the Inupiaq's familiar caribou-hunting and egg-collecting grounds shown in stark relief. As the film ran a slick and oily voice-over intoned: "Consider the development of a harbor at a remote, coastal area. The procedure: Bury an inline series of four, equally spaced 100-kiloton shots and a terminal shot of one megaton. *Detonate!*"

And at the word *"Detonate!"* the area erupted in a blinding white flash and the ground heaved up and exploded and fanned out in the classic, circular, crown-shaped pattern of a large-scale nuclear bomb blast—to the yelps, cries, screams, wailing, and shrieks of horror from the assembled Eskimos as they saw their very own ancestral hunting grounds blown to smithereens and lofted up to high heaven.

Well, this was a public relations disaster of the first order, since the film had the precise opposite effect from that which had been naively intended. That lesson was bolstered later on in the meeting when one of the Eskimos, Kitty Kinneeveauk, got up and announced to the audience and speakers in her somewhat halting English: "So, I've been thinking about we really don't want to see the Cape Thompson blasted because it our homeland. I'm pretty sure you don't like to see your home blasted by some other people who didn't live in your place like we live in Point Hope."

Which, as it turned out, pretty well reflected the prevailing mood among the Inupiaq in attendance. At the end of the meeting, after the AEC reps said bye-bye to the villagers and flew away in their aircraft, the members of the Point Hope Village Council voted on Project Chariot. To no one's surprise, the result was a resounding and unanimous "No."

Following the Point Hope public relations debacle, the AEC and Plowshare officialdom took stock of their situation to decide whether to continue with Project Chariot, with the explosion now scheduled for the spring of 1962. They chose to beat on, boats against the current . . . but a series of unforeseen events was soon put in motion that in the end would sink the whole effort.

In June 1961, Barry Commoner's Committee for Nuclear Information (CNI), a group located at Washington University in St. Louis, published a special issue of their newsletter, *Nuclear Information*, devoted solely to the risks and potential benefits of Project Chariot. The twenty-page report was scientifically well informed but also generally hostile to Chariot and to what it regarded as the AEC's overoptimistic claims about the safety of the blasts. The authors emphasized the difficulty of predicting wind conditions in the area, and said that in consequence there would be considerable uncertainty about the direction of travel of the fallout cloud.

The CNI report also zeroed in on an oddball problem concerning the caribou. Caribou were a staple of the Eskimo diet, and they were plentiful in the Cape Thompson region. The animals had a particular affinity for lichens, which were themselves weird fungal-algal combination plants that got their water not from the ground, as per usual for a plant, but from the atmosphere. As such, the plants were sensitive indicators of atmospheric pollution levels. It turned out that when the University of Alaska biological team was studying Cape Thompson plant life, they discovered that the area's lichens contained high levels of the radioactive element strontium 90, as did the caribou that feasted on them. Lichens soaked up the element from the ambient fallout already present in the atmosphere, and the caribou then picked it up in turn as they ate. Even though Alaska had generally low levels of fallout, the area's caribou had about the same levels of strontium 90 in their bones as the practically mutant, glowing, radioactive cattle raised in the vicinity of the Nevada Test Site. Since the Eskimos ate caribou meat, they too would be at risk for accumulating high levels of strontium 90 in their bones and elsewhere in their bodies.

The day after the CNI report appeared the *New York Times* ran a story headlined "Caribou May Bar Alaska A-Blast." The problem was not that the blast would harm caribou—the problem was that caribou meat would harm the Eskimos who consumed it. "In the process of blasting out the harbor," the *Times* story said, "the [Atomic Energy] Commission might contaminate the food chain in the Arctic region so that radioactive strontium would pass from plants into animals and thence into the bones of Eskimos."

The impact of the CNI account was blunted somewhat by a critical piece in *Science*, but the CNI authors responded with a point-by-point rebuttal that the magazine also published. Between the *Times* and the *Science* stories, the outside world had finally been informed of the strange nuclear goings-on that were afoot, or were about to be, in the far-off new state of Alaska.

Finally, in April 1961, *Harper's* magazine delivered the coup de grâce in an article entitled "The Disturbing Story of Project Chariot." The piece laid out the glorious history of the project— then in its fourth year—and showed how the original, ambitious plan for an economically self-supporting commercial harbor had metamorphosed into a proposal that amounted to little more than a set of blasts aimed at conducting a proof-of-concept *experiment* that might pave the way for the Central American canal project. "The essence of an experiment," the authors noted, "is uncertainty." And there was plenty of uncertainty about the effects of the proposed nuclear detonations and the radiation hazard they presented to the area's plant, animal, marine, and human life.

The *Harper's* article mobilized public opinion and rattled many of the project's supporters. By the end of that April, the Livermore scientists recommended that the AEC cancel Project Chariot once and for all. Their rationale was that the winds at the site were quite changeable and could not be guaranteed to blow in the right direction, either at the time of the blast or immediately afterward. In addition, the results obtained from an explosion in permafrost could not reliably be extrapolated to crater wall stability elsewhere, such as in the Central American tropics, the site of the proposed Pan-Atomic Canal. Plus, other less public and potentially less controversial nuclear tests could provide much of the same information as that which had been expected to be derived from the Chariot blasts.

And so, after several months of delay, the AEC officially canceled Project Chariot. But not Project Plowshare. The atomic scientists could simply not let go of their pet notion of performing miraculous geographical engineering feats by means of nuclear bombs.

The reason they couldn't is bound up with the emotional hold, the spell, the enthrallment that large-scale releases of nuclear energy had for the scientists. The atom had long been viewed as an indivisible, indestructible, and eternal object, impervious to breakup or other forms of alteration. Breaking up the atom (as took place in a fission bomb) and fusing atomic nuclei together (as happened in a fusion bomb), together with the consequent release of enormous amounts of energy, were two of the epochal feats ever performed by science. Those who witnessed the detonation of the first fission bomb, the "gadget," at Trinity Site in the New Mexico desert knew at once that they were in the grip of something that was truly awesome, overwhelming, mesmerizing.

Many of the reactions of those present at the Trinity Test have passed into history, and at least one has been quoted so often as to have become a cliché. This was the line from the Bhagavad Gita that passed through Robert Oppenheimer's head when he finally saw the explosion that he had done so much to produce: "Now, I have become death, the destroyer of worlds." Lesser-known responses to the blast also betrayed evidence that the beholders had been in the presence of something truly extraordinary, something that transcended the limits and forms of daily living, and in fact of all ordinary experience.

"No one who saw it could forget it," said Kenneth Bainbridge, director of the Trinity test. "A foul and awesome display."

"The grandeur and magnitude of the phenomenon were completely breathtaking," said Robert Serber, a Manhattan Project scientist.

"The whole spectacle was so tremendous and one might almost say fantastic that the immediate reaction of the watchers was one of awe rather than excitement," said Edwin McMillan, a Los Alamos physicist. "I am sure that all who witnessed this test went away with a profound feeling that they had seen one of the great events of history."

But perhaps the description that best conveyed the intense emotional quality and psychological effect of the blast came from Brigadier General Thomas F. Farrell, who was in the control shelter 10,000 yards south of the detonation point, together with Oppenheimer, Bainbridge, George Kistiakowsky, and several other observers who had been instrumental in creating The Bomb. Farrell wrote: "The effects could be called unprecedented, magnificent, beautiful, stupendous and terrifying. No man-made phenomenon of such tremendous power had ever occurred before. The lighting effects beggared description. The whole country was lighted by a searing light with the intensity many times that of the midday sun. It was golden, purple, violet, gray and blue. It lighted every peak, crevasse, and ridge of the nearby mountain range with a clarity and beauty that cannot be described but must be seen to be imagined. It was the beauty the great poets dream about but describe most poorly and inadequately. Thirty seconds after the explosion came first, the air blast pressing hard against the people and things, to be followed almost immediately by the strong, sustained, awesome roar which warned of doomsday and made us feel that we puny things were blasphemous to dare tamper with the forces heretofore reserved to The Almighty. Words are inadequate tools for the job of acquainting those not present with the physical, mental and psychological effects. It had to be witnessed to be realized."

That momentous display and all of its physical, emotional, and psychological by-products marked the beginning of the Atomic Age. How could any of those who were involved in it and who witnessed its birth and formative years not want to harness the metaphysically transcendent power of the atom? And not merely in the rather pedestrian form of nuclear power generation, which, compared to an atomic bomb blast, was an utterly plain and boring technical performance. Especially for those who, like Teller, had helped invent the far more powerful "Super," the H-bomb, there was nothing that could compare with the barely controlled violence of a thermonuclear explosion. Here was a force, Edward Teller thought, that practically cried out to be harnessed and used; it was one that could in fact "work miracles."

And so Teller and the rest of the Plowshare fraternity now tried all over again to start performing them. Across the seventeen-year course of the Plowshare program, the atomic scientists considered projects for canals, harbors (including one in Hawaii), reservoirs, river diversions, and railroad and highway cuts, in locations as diverse as the Philippines, Thailand, Greenland, Madagascar, India, Pakistan, South Korea, and Twin Falls, Idaho, among other places.

But in the immediate wake of the Project Chariot cancellation, the Plowshare scientists merely retreated to the Nevada Test Site, where they proceeded to blow some extra-large holes in the ground in preparation for blowing an even bigger series of them across the whole of Central America. The biggest experimental blast was named Project Sedan (the logo for which was a sedan chair), which was a single-shot, 100-kiloton cratering test that did not have any obvious relationship to the Central American canal project, since it would provide data neither about the effects of a row-charge explosion nor about the action of an explosion below water. It was a test, furthermore, after which practically everything that could possibly go wrong did.

The hydrogen bomb went off, as planned, at 10:00 a.m. on July 6, 1962, and as predicted, it made a very big crater—1,200 feet across by 320 feet deep—and released a wave of seismic energy equivalent to an earthquake with a magnitude of 4.75 on the Richter scale. The AEC, in a press release the next day, pronounced this abyss, and the tons of dirt and rock that had been ejected from it, as having made a "significant contribution to nuclear earth-moving technology." But that claim ignored the collateral damage.

Veterinarians from the US Army Veterinary Corps had placed "biological air samplers" (i.e., "animals") at monitoring stations located thirty-one miles and forty-two miles from "GZ" (ground zero). As noted in their final report on the iodine inhalation study, "Purebred beagle dogs were used as biological air samplers. These animals were obtained several weeks prior to the operation and were housed in kennels at the headquarters compound. To quantitate air sampled by the dogs, and to relate the radioiodine collected with that collected by the physical air sampling systems, respiratory frequency and inspired air volume of each animal were determined before the Sedan event."

At specified periods after the test, the beagles were "sacrificed" (killed) and their various organs were examined to determine the "burden of gamma emitters," which was to say, the quantities of radioactive isotopes that had been collected inside them. The amounts found were much higher than expected, and in the case of the large intestine and its contents, the radioactive activity "exceeded counting capacity."

The dust cloud from the Sedan explosion proved to be 50 percent bigger than anticipated, and the fallout turned out to be five times greater than forecast. Some 250 tons of radioactive dust were lofted into the atmosphere and drifted across Nevada, over into Utah, and then across the rest of the country on its way to the Atlantic. The cloud of debris was so thick when it passed over Ely, Nevada, that the streetlights winked on. The fallout deposited on cattle grazing land was eaten by the animals, whose radioactive milk was later consumed by humans, including infants. And so on down an extended chain of other unintended side effects.

In the classic manner of the practitioners of a pathological technology, AEC spokesmen scoffed at the idea that any of these consequences posed dangers to plants, animals, or human beings. Most of the radioactivity, they claimed, fell "close to the crater and within the test site." The Sedan test, they said in a news release, showed "that earth-moving projects of this kind can be conducted, under controlled conditions, in safety."

In September 1964, more than two years after the Sedan shot, the US Congress appropriated $17.5 million for a feasibility study of the Central American Pan-Atomic Canal. In December of that year, the new president, Lyndon Baines Johnson, announced that nuclear excavation was to be "national policy" and said of the canal project, "This new age requires new arrangements."

But several factors spoke against the nuclear excavation of the new canal. For one thing, even so much as finding all those people living in the so-called exclusion zone was going to be much harder than anyone had at first anticipated. "These Indians are unfriendly to outsiders," reported one investigator, who felt that the native Kuna Indians, in particular, might well "resist attempts to evacuate them and undoubtedly many will hide in the impenetrable jungle nearby. Some who are not found and forcibly evacuated would be killed or injured by the nuclear explosions."

As for those who were found and relocated—which could be as many as 30,000 people—provision would have to be made for them to live elsewhere, perhaps for many years, after which it was doubtful whether many of them would want to return to their once-radioactive homes. In any case, this forced mass migration of local residents would be inordinately expensive, with the whole project conceivably ending up costing even more than digging out a new canal by conventional, non-nuclear means.

In 1965 John Gofman, head of Livermore's biomedical division, bluntly informed his bosses: "My view is that building a canal with hydrogen bombs would be biological insanity."

On December 1, 1970, the Canal Study Commission formally reported to the next new president, Richard Nixon, that, "unfortunately, neither the technical feasibility nor the international acceptability of such an application of nuclear excavation has been established at this time." The canal project, finally, was dead.

And by the mid-1970s so was Project Plowshare, its overhyped "miracles" remaining unperformed, at least by nuclear detonations.

The Tenn-Tom Waterway, a 234-mile man-made marine transportation channel, would be completed in 1984 by conventional means, and despite the fact that it was the largest earthmoving project in history, it needed no hydrogen bombs for its construction.

The exact date of Plowshare's formal termination is somewhat obscure. Scott Kaufman, author of the recent book *Project Plowshare,* noted that the date "when Plowshare officially died is a matter of debate." However, a public information sheet released by the National Nuclear Security Administration in August 2013 describes Plowshare as having continued "until 1975," while a US Department of Energy document gives June 30, 1975, as the date of the project's formal termination.

But Plowshare ended only after its scientists had conducted a long series of nuclear, as well as non-nuclear, explosions under its auspices. Starting with the nuclear test Gnome, in New Mexico in 1961, and ending with a grand simultaneous triple nuclear detonation in Colorado in 1973 (Rio Blanco shots 1, 2, and 3), the United States set off a total of thirty-five separate underground nuclear devices as integral parts of Project Plowshare. The project also included twenty non-nuclear tests conducted over roughly the same twelve-year period, making for a total of fifty-five explosions.

Project Plowshare was therefore no small-time effort. In the scale of its radiation-producing effects and the magnitude of its ambitions to undertake massive nuclear engineering feats of various different types in the United States and across the globe, Plowshare represented a high point in the annals of pathological technology. It was huge in its scope and aspirations, it was motivated by the emotive vision of turning nuclear bombs into agents of good, it systematically minimized the risks involved, and its benefits were outweighed by its likely costs to human life, health, and the environment.

As the Department of Energy's report on the project itself conceded, "Plowshare was a program that started with great expectations and high hopes. Many engineering projects did not progress
beyond their planning phase and construction was not started. In
general, planners were confident that the projects could be completed safely, at least within the guidelines at the times. There
was less confidence that they could be completed cheaper than
by conventional means and most importantly, there was insuffi-
cient public or Congressional support for the projects."

In this case, Congress and the American public showed far
better judgment than the nation's best nuclear scientists, who had
their jobs to protect and their own axes to grind. Project Plowshare was thus a striking practical example of the truism that
some decisions are too important to be left to the experts.

But in this regard, Plowshare technology was by no means
alone. The world's top scientists were not always reliable judges of
a given technology, even when its use was to further science itself.

Chapter 10

THE GODZILLA
OF PHYSICS

To be classified as pathological, a technology does not have to kill people, maim them, or subject them to any kind of physical risk. A nonlethal technology that is not physically harmful may nevertheless fully satisfy the four criteria of what makes a technology pathological. The Superconducting Supercollider (SSC), a gigantic particle accelerator that was to have been built in Texas during the 1980s and 1990s, is an example of such a technology.

Until roughly the beginning of the twentieth century, practically all branches of science, including physics, astronomy, chemistry, biology, and even microbiology, were primarily concerned with the easily observable realities of ordinary experience. Some of these items were of such a nature that scientists had to use instruments in order to observe them: telescopes for faraway planets, stars, and galaxies, and microscopes for smaller organisms such as bacteria and tiny structures such as the chromosomes. But all these entities were nevertheless ordinary, macro-scale, visible objects. Some were larger than the familiar artifacts of everyday life, some were farther away from us, and at the other extreme some were far smaller than the mundane things that we interact and deal with on a daily basis.

For much of the history of science, this has been true even for the phenomena studied by physicists. Before the twentieth century, physics was concerned with quite ordinary, everyday realities: light, heat, and sound; motion, acceleration, and falling bodies; gases, fluids, and solids; and electricity, magnetism, and so on and so forth through the world of common experience. Then Wilhelm Roentgen discovered X-rays in 1895, J. J. Thomson discovered the electron in 1897, and Ernest Rutherford discovered the proton in 1914—and all at once a new branch of physics had come into existence: elementary particle theory, a science that dealt with the hidden and the invisible, the fundamental entities underlying the observed phenomena of everyday life.

These new entities were radically different from the objects that normally could be seen and touched. Indeed, these creatures of the invisible underworld were so far removed from daily life, so far below the level of direct, or even indirect, sensory perception, that they might as well have been off in another realm. Which, in a sense, they were: subatomic particles were not governed by the same laws as those that held for macroscopic bodies; rather, they operated by their own special rules, the principles of quantum mechanics, which had no obvious and intuitive relationship to the laws of day-to-day experience. Years of education, training, experimentation, and apprenticeship were required to become familiar with these arcane particles and to comprehend their idiosyncrasies—to the extent that they could be "comprehended" at all. For even experts in the field, masters of these nether re- gions, themselves readily acknowledged their bafflement at the laws governing the occult phenomena in which they specialized. Nobel Prize–winning particle physicist Steven Weinberg, for example, confessed in print: "I admit to some discomfort in working all my life in a theoretical framework that no one fully understands." Or as Richard Feynman, America's physicist-in-chief and one of the inventors of quantum electrodynamics, put it in his pithy little primer *The Character of Physical Law:* "I think I can safely say that nobody understands quantum mechanics. . . . Do not keep saying to yourself, if you can possibly avoid it, 'But how can it be like that?' because you will get 'down the drain,' into a blind alley from which nobody has yet escaped. Nobody knows how it can be like that."

To the extent that quantum mechanics *was* actually understood, knowledge of its principles, entities, fields, and forces had early on become the private preserve of a small, elite group of physicists known as elementary particle theorists. Given the nature of the objects with which they dealt, and the fact that the particles existed off in their own special realm, it was entirely understandable that this group of theorists had as much as become a priesthood—and a tiny one at that. Theirs was a far smaller group than any conference of bishops that presided over other major religious denominations, except perhaps the few remaining Druids. This particle physics priesthood had its own remote subject matter, as well as its own arcane language, set of symbols, methodologies, ceremonies, rules, and rituals. It was a religion whose practitioners talked to, and were understood by, each other and nobody else.

These wizards of the quantum realm formed a special brotherhood (which included a few sisters too) who were privy to exclusive, cabalistic, almost forbidden knowledge. They had a distinct way of looking at the world, a specific new way of thinking, talking, and theorizing about nature. Their Weltanschauung was captured in myriad recondite symbols, equations, formulas, and constants, plus entire new vocabularies that referred to whole new populations of things: quarks, leptons, baryons, muons, gluons, mesons, kaons, pions, photons, bosons, alpha, beta, and gamma rays, Ω particles, π particles, W^+ particles and W^- particles, Z particles and even Z^0 particles, neutrinos, antineutrinos, hadrons, fermions, partons (by now long forgotten), positrons, spins, lattices, symmetries, strong, weak, and electroweak interactions, superstrings—and much more besides.

This zoo of exotic particles, fields, and forces was unfamiliar and foreign to the average layperson, but that was only to be expected given that the phenomena in question resided at the most subterranean and hidden levels of the material universe. For that matter, the basic entities of molecular biology were almost as alien, and it would be easy to compile a list of comparably abstruse molecular-biological entities: DNA, RNA, mRNA, tRNA, rRNA, stop codons, start codons, introns, exons, promoters, and so on. There was an important difference between the two lists, however. The elementary particles of physics were largely of explanatory or theoretical value: they helped scientists understand the nature of matter, but most of these entities had little obvious practical value. The objects of molecular biology, by contrast, had enormous practical and therapeutic significance: they could be manipulated and changed in order to improve human life and health and indeed were so utilized in medicine.

There was a further difference in the scale of the scientific instruments that were necessary to discover, understand, and, in the case of molecular biology, utilize the objects of the two disciplines. Molecular biology was essentially small science: you could perform most of its operations with machines that fit on a desktop—gene sequencers, gene synthesizers, PCR (polymerase chain reaction) machines, and the like. But high-energy physicists operated on a far grander scale: theirs was the province of truly immense machines, of particle accelerators that were miles in length or tens of miles in circumference. These instruments accelerated invisible subatomic bodies to velocities approaching the speed of light, and then smashed them into a target, or each other, and observed the results, which appeared in the form of esoteric images expressed in sets of crabbed, cryptic, tea-leaf-like markings rendered on photographic plates, giving a picture of the subatomic wreckage that has been likened to an explosion in a piano factory.

Truly, particle accelerators were the cathedrals of the new religion. This was stated openly and in as many words by the masters of the realm themselves. For example, Robert Wilson, the director of Fermilab, the four-mile-long circular accelerator at Batavia, Illinois, near Chicago, said: "I even found, emphatically, a strange similarity between the cathedral and the accelerator: The one structure was intended to reach soaring heights in space; the other is intended to reach a comparable height in energy. Certainly the aesthetic appeal of both structures is primarily technical. In the cathedral we see it in the functionality of the ogival [pointed] arch construction, the thrust and then the counterthrust so vividly and beautifully expressed, so dramatically used. There is a technological aesthetic in the accelerator, too. There is a spirality of the orbits. There is an electrical thrust and a magnetic counterthrust. Both work in an ever upward surge of focus and function until the ultimate expression is achieved, but this time in the energy of a shining stream of particles."

One might view this as nothing more than a metaphorical, almost poetic, expression of certain commonalities between the

two structures rather than a statement to be taken literally or at
face value. But then Leon Lederman, Wilson's immediate successor as director of Fermilab and a Nobel Prize–winning physicist, only reinforced the same point when he said, "There is this
deeper connection. Both cathedrals and accelerators are built at
great expense as a matter of faith. Both provide spiritual uplift,
transcendence, and, prayerfully, revelation. . . . Accelerators are
our pyramids, our Stonehenge."

But even Fermilab, cathedral-doppelgänger that it was, proved
not to be the ultimate in particle accelerators. Leon Lederman
and Steven Weinberg were the two main leaders of the effort
to build the latest and greatest of the genre, a Brobdingnagian
creation called the Superconducting Supercollider (SSC). Its
purpose was to create yet one more new subatomic particle, a
never-before-seen but much-theorized-about entity lurking at the
bottom of nature, at the innermost recesses of matter: the Higgs
boson.

It was Lederman himself who bestowed upon the Higgs boson
the name that was forever afterward associated with it, for better
or for worse: the God Particle.

As a sub-branch of physics, the origins of particle theory were
humble enough. It began in 1897, when the British physicist
J. J. Thomson, director of the Cavendish Laboratory at Cambridge
University, discovered the electron, the first subatomic particle to
be produced and observed experimentally. He did this by means
of what was in effect a primitive particle accelerator. The device
was a cathode ray tube—a closed, evacuated chamber about the
size and shape of a wine bottle lying on its side. At the cork end
of the bottle, a cathode, a negatively charged electrode, emitted a
stream of negatively charged particles. A little farther up the tube,

about where the neck of the bottle flared out, was a disk-shaped anode, a positively charged electrode with a hole in the center to let the particle stream pass through. The anode attracted the streaming particles and thus sped them up. By any reasonable standard, then, Thomson's device was in fact a genuine, working particle accelerator.

At the far end (bottom) of the bottle, the stream of charged particles caused the glass to glow or fluoresce, in a process similar to that which would produce the image on a computer screen or a television (one of the old-fashioned cathode-ray-tube variety). What was important about Thomson's experiment, however, was not the glow that appeared at the end but rather what happened to the particle stream as it passed between a separate set of charged plates positioned at the *sides* of the bottle. One held a positive charge, the other negative.

By varying the voltage on the two side plates, Thomson was able to deflect the beam (which was visible as a narrow shaft of light) upward and downward by different amounts. And by measuring the strength of the applied voltages as against the amount of deflection that resulted, he calculated that the stream had to consist of exceptionally light, negatively charged particles— particles that had far less mass than even the smallest atom, hydrogen.

"I can see no escape from the conclusion that [cathode rays] are charges of negative electricity carried by particles of matter," he said. Such negatively charged particles, Thomson believed, constituted "a state of matter more finely subdivided than the atom."

Thomson's cathode-ray-tube accelerator was undoubtedly one of the cheapest such devices in the history of particle physics, probably costing not much more than a bottle or two of good wine at the time.

Discovery of the proton, on the other hand, did not even require an accelerator. It was found in 1914 by the New Zealand physicist Ernest Rutherford, who had worked for a time as assistant to J. J. Thomson at the Cavendish lab. Rutherford's experiment required only a source of radiation and a piece of gold foil. At Manchester University, Rutherford, together with postdoc Hans Geiger (of Geiger counter fame) and assistant Ernest Marsden, shot a stream of alpha particles (helium nuclei) from a radium sample toward a thin sheet of gold metal.

Rutherford had expected the alpha particles to pass through the gold atoms without resistance, since he viewed atoms as amorphous blobs of matter studded here and there by insubstantial electrons—the old "raisin pudding" concept of the atom. To his surprise, however, he observed periodic flashes of light, as if a few of the alpha particles were actually bouncing back off the foil, which to him was a totally unexpected result. "It was almost as incredible as if you fired a 15-inch shell at a piece of tissue paper and it came back and hit you," Rutherford said later.

He viewed this outcome as evidence that atoms had hard, nuggetlike centers and were not at all like permeable, homogeneous raisin-pudding blobs, as had been thought. His discovery was the basis of the new "Rutherford atom," modeled ever since as a small solar system with the central, sunlike nucleus surrounded by one or more orbiting electrons. Finally, with James Chadwick's 1932 discovery of the neutron, which also resided in the nucleus, the gross description of the atoms that constituted ordinary matter was essentially complete. The structure of the Rutherford atom was simple, clear, and intuitively plausible, even if it was not literally or entirely correct. Still, even a child could understand it.

This, of course, did not mean that there were no more mysteries, problems, or issues left dangling regarding the atom or its component parts. The most obvious question remaining was whether the three subatomic particles were themselves rock-bottom entities, irreducible and not themselves made up of anything smaller or more fundamental. Another obvious question—whether anything else existed besides electrons, protons, and neutrons—had already been answered in the affirmative by Roentgen in 1895, with his discovery of X-rays, which were a form of electromagnetic radiation. So what else was out there? Or more precisely, what else was *in there,* deep down inside the atom?

And there were other puzzles—such as that of beta decay, a radioactive process in which electrons were emitted from the nucleus. That was a puzzle because there *were* no electrons in the nucleus. So how could they *exit* the nucleus without being there in the first place? (The answer was that they were created in the process of a neutron's disintegration, an event that also gave rise to a proton. Yes, well, just don't ask: "How can it be like that?")

Such questions and conundrums were at the heart of fundamental particle theory, the attempt to discover the most basic and encompassing laws of nature. But to find the answers, physicists had to break up the atom and its component parts, not with dynamite or sledge hammers, but rather with particle accelerators, which sped up electrons and protons to near-relativistic velocities and crashed them into prepared targets—or in the case of colliders, into each other. It may have been an odd way to acquire new knowledge about nature, but it worked.

The simplest particle accelerator is a straight evacuated tube plus a mechanism for applying electrical voltages. But a linear accelerator can get only so long before its size makes it unmanageable. Thus, it was a milestone of particle accelerator development when Ernest O. Lawrence invented the cyclotron,

a circular accelerator. (This was the same Ernest Lawrence who
would later be one of the founders of Project Plowshare.) The
chief advantages of a cyclotron were that you could confine particles to smaller spaces than with linear accelerators, and in addition you could impart progressively stronger electrical kicks to
them, and thus increase their speed, each time they went around
the racetrack.

The genesis of the device is the stuff of myth, but according to
the received account, Lawrence came up with the idea in 1929,
when he was an associate professor of physics at the University of
California at Berkeley, after perusing an article by the Norwegian
Rolf Wideröe in a German electronics journal, *Archiv für Elektro-
technik.* Lawrence, who was from South Dakota, wasn't a fluent
reader of German and did not in fact read the text: "I merely
looked at the diagrams and photographs of apparatus," as he later
recalled. Nevertheless, "simple calculations showed that the [lin-
ear] accelerator tube would be some meters in length which at
that time seemed rather awkwardly long for laboratory purposes."

It would make more sense, he thought, to bend the apparatus
back in upon itself so that the particles would be forced to go
around in circles (more precisely, in progressively wider spirals),
while timed electrical impulses were applied to them on each
pass. Others who have told the story have added the embellish-
ments that while crossing the campus from the library to his office Lawrence encountered a faculty member's wife, to whom he
blurted out, "I'm going to be famous!" and that he later sketched
out his basic design on the canonical paper napkin. Whatever the
truth of these accretions, Lawrence's device undeniably consisted
of two semicircular accelerating chambers ("dees") separated by a
small gap. Particles introduced into one of the dees were attracted
toward the other one by a difference of electrical potential, and

then conversely attracted back again by a reversal of the voltage difference. In all of their travels, the particles would be kept in a circular path by the force fields created by an adroitly manipulated system of external magnets.

Conceptually, then, the cyclotron was simple enough, its two main features being electrodes to accelerate the particles—operating like a wave of electrical energy pushing a surfer along—and a series of magnets to hold the particles in their orbits. Lawrence constructed a prototype out of random scraps of metal, wire, glass, and sealing wax, among other things, and jokingly referred to the result, which was a mere four inches across and had cost all of $25, as "a proton merry-go-round." He later christened it a "cyclotron."

The four-inch device was followed soon enough by a nine-inch model, and then an eleven-inch machine, and so on, after which cyclotrons just kept getting bigger and bulkier until, despite his original intention of keeping them inside the laboratory, Lawrence's "atom smashers" had outgrown the lab to such an extent that they had to be moved into their own separate building. Thus, the fact that they were circular rather than linear didn't mean that they couldn't become king-size, even oversize, devices, as well as hugely expensive. With increased size, of course, came the need for higher energies to operate them.

In addition, the bigger these atom smashers got, the smaller and more evanescent were the pieces of subatomic detritus they produced, and the farther away these entities receded from ordinary experience or intuitive understanding. Progressively, these inscrutable, otherworldly particles were of interest to, and comprehensible by, only the high priesthood of high-energy particle theory plus a few die-hard science and technology junkies, and very few others.

And then, at a certain point, particle accelerators crossed a dividing line. Not only did they outgrow the lab, and then the separate buildings that had contained them, but they eventually expanded to such a gigantic, enormous, out-of-control jumbo size that they were big enough to enclose not only one building but several buildings, and then even small villages. At that point, their dimensions were no longer measured in inches, feet, yards, or furlongs, but rather in miles.

It was ironic that whereas Lawrence's initial motivation in creating the cyclotron was to make the device smaller, once they had been created these machines did nothing but get bigger . . . and bigger . . . and bigger. The question then arose whether there was an upper limit to the size of an accelerator. The question was not so much whether there was an upper *physical* limit—as it was clear that accelerators could be made arbitrarily large—but rather whether there was an upper *functional* limit, a limit established by the prospect and then the reality of diminishing returns in proportion to the machine's increased size, expense, and effective and intelligent utilization of resources.

In the United States, the advance toward the upper limits culminated, at least temporarily, in 1967 with the Atomic Energy Commission's creation of Fermilab, Robert Wilson's accelerator in northern Illinois, west of Chicago. Its main ring was to be four miles in circumference and more than a mile across. The full laboratory facility consisted of more than 100 buildings, approximately one million tons of infrastructure, instrumentation, and equipment, and more than 1,500 employees. The accelerator's main ring itself enclosed more than ten square miles of area, which included a lake, woods, and flat grassland on which a herd of buffalo roamed, plus deer, fox, raccoons, geese, ducks, about forty species of birds, and assorted other wildlife. Along with, of course, the particle physicists themselves, plus their grad students and post-docs.

Fermilab did not in fact encircle a village, but in order to be built, it had displaced one out of existence. The village was Weston, a small town in Kane County, Illinois, that had been established at the end of World War II. Years later, in the mid-1960s, AEC officials were considering a roster of places that they had whittled down from more than 400 to some 85 sites as possible locations for their new National Accelerator Laboratory. Then they reduced the list to six. On April 8, 1966, a group of AEC reps visited the town of Weston, one of the six finalists, which greeted them with a line of fire trucks, a color guard, and a marching band. On December 16 the AEC finally announced that Weston had been chosen as the site.

But there was a catch. In order for the accelerator to be located at Weston, the town's total population would have to pick up, depart from, and say good-bye to their fair city forevermore. According to Fermilab's own web page, "This was a part of the legal necessity to properly turn over the entire site to the US Atomic Energy Commission by the state of Illinois, which had purchased the entire acreage." And so, on Wednesday night, November 26, 1969, three years after the site selection—during which time the city had erected a sign proudly announcing: WELCOME TO WESTON, FUTURE ATOMIC RESEARCH CAPITAL—the village board met for the last time and voted to approve an ordinance to vacate the town. Before the meeting was adjourned, Lewis V. Morgan, a state representative, hailed the turning over of the village to the National Accelerator Laboratory as "the scientific gift of the century."

The village of Weston was thereupon officially dissolved, its population (which was now zero) disappeared from census figures, and the town similarly vanished from maps of Illinois. After the accelerator was constructed, it did in fact discover some new particles, including the top quark, the tau neutrino, and the bottom Omega baryon. Nor was the project fantastically expensive: when completed, the facility had cost a total of $243 million, which was even slightly under budget.

But that cost savings had come at a price. The main ring's primary operational structures were a series of 1,014 magnets that kept the thin proton beam circling around properly within the boundaries of its prescribed path. The magnets were laid end to end, around the full four miles of the ring's circumference. In 1971, just as the machine was in the process of getting up and running, the magnets began to explode. "They blew excruciatingly, one magnet at a time," according to "The Building of Fermilab" by science writer Philip J. Hilts. When one magnet failed, the entire four-mile-long ring of them stopped functioning. It was like a set of Christmas tree lights wired in series rather than in parallel, so that if one of the lights blows, the whole string of them fails and all blink off simultaneously.

It turned out that many of the magnets had hairline cracks in their fiberglass insulation covering. Before the problem was solved, more than 350 of them had failed—roughly one-third of the total—over the course of almost a year's worth of trials, fixes, and restarts. But by March 1972, at long last, all of the cyclotron's 1,104 magnets were finally working concurrently.

Still, even this four-mile-long accelerator loop, with all of its exploding magnets, speeding protons, and frazzled baryons and quarks, was not quite oversize enough to get all the way down to the rock bottom of physical existence, to the all-time lowest depths of matter's "ultimate nature." Obviously, to plumb the *ultimate nature* of matter you needed to build the *ultimate cyclotron*. And so, of course, the nation's high priests of high-energy physics proposed to do just that.

The ultimate cyclotron would be the biggest machine of any type ever constructed. It would be the Godzilla of physics: the Superconducting Supercollider.

B ecause the Superconducting Supercollider (SSC) was abandoned before it was completed, it is difficult to say precisely what its "final" design was. Specifications for the machine and its components changed over time, and no two descriptions of the device seem to agree exactly on any one parameter. The two principal and most vocal proponents of the SSC were both Nobel Prize–winning high-energy physicists—Steven Weinberg and Leon Lederman—and as it happened, each of them had written (or cowritten) a book touting the virtues and insisting on the scientific necessity of their planned megadevice. But in a tour de force of bad timing, both books came out in 1993, the very year in which the US Congress terminated funding for the project once and for all.

Nobody in Congress is on record as having branded the SSC an example of a pathological technology, but the instrument as proposed arguably satisfied all four of the relevant criteria. First, *that the machine was to be huge was clear on the face of it*: the device was to be fifty-four miles in circumference, which was about as long as the Washington, DC, Beltway, longer than two marathons run back to back, and longer even than the Panama Canal. Indeed, its dimensions would make the SSC into the biggest scientific instrument of any kind ever built. (The Large Hadron Collider [LHC] operated by the European Organization for Nuclear Research [CERN] in Switzerland, at seventeen miles around, was less than one-third the size of the SSC.)

The device would require the excavation of some ten million cubic feet of earth, after which every last mile of the resulting tunnel would be filled with a series of heavy, but delicate, and very expensive mechanisms: a circular array of some 10,000 electro-magnets, each of which would be fifty-five feet long and weigh approximately twelve tons. These magnets were to be supercon-ducting, meaning that they would offer zero electrical resistance. Such devices would allow for higher energies and more efficient operation than conventional electromagnets since no energy would be lost as waste heat. But to achieve superconductivity along the main ring's fifty-four miles of length would require a lot of cooling: to wit, two million liters of liquid helium, refrigerated to minus 452.2 degrees Fahrenheit (minus 269 degrees Celsius), or about four degrees above absolute zero, throughout the entire length of the structure. This in and of itself was such a vast amount of re-frigeration that it more than doubled the amount of liquid helium on all of planet Earth. And as zeppelin builders knew all too well, helium was not cheap even at room temperature.

The rest of the SSC, which was to be built around the city of Waxahachie, Texas, was comparably heroic in scale. Indeed, the entire complex was so oversize that it held the rare distinction of making the *Hindenburg* look microscopic. So when it came to satisfying the first criterion of a pathological technology, the SSC was, if anything, an overachiever.

Second, in carrying on the grand and historic tradition of at-tempting to understand all of nature and to plumb the innermost secrets of the material universe—a search that went back to the very first scientist-philosophers of ancient Greece, if not before—*the enterprise was firmly rooted in emotion.* High-energy physicists were eloquent in expressing the powerfully stirring nature of their quest. As Robert Wilson had said of Fermilab, "Our work here is primarily spiritual. We are concerned with the ultimate na-ture of matter." It was mildly surprising to hear a *physicist,* of all people, equate the spiritual and the material, but Wilson's point nonetheless came across quite clearly: their work was so momentous as to be of a transcendental nature.

Leon Lederman, for his part, said that with the SSC "the world of science was on the brink of a series of exciting new discoveries that would bring us closer to understanding how the universe works and the identities of the building blocks that make it possible." And further, "Behind the God particle is revealed a world of splendid, blinding beauty, but one to which the mind's eye will adapt." It was almost a vision of heaven.

In the face of such uplifting grandiloquence, Steven Weinberg was practically down to earth when he spoke of "the great intellectual adventure of discovering the final laws of nature" and claimed that the "final theory" was "like a piece of fine porcelain that cannot be warped without shattering." (He also alleged that "opposition to the SSC" stemmed from a "mindless distaste for any large new technological project," implying that a reasoned, principled, or intelligent opposition to it was barely even conceivable.)

Third, *supporters of the SSC tended to minimize its risks or, worse, did not even acknowledge the existence of any riskiness to the design, construction, operation, or final success of the machine.* But there were risks aplenty to the SSC, the first and most basic of which was that it wouldn't discover the Higgs boson after all. The machine had been designed to perform an experiment, and the results of any experiment were by definition unknown beforehand. A favorable outcome was anything but guaranteed. Indeed, the whole project was a leap of faith, as Lederman had said. Freeman Dyson, himself a mathematical physicist, agreed, adding that "every big new machine is a gamble. If the gamble succeeds, the machine produces important new discoveries. If the gamble fails, the machine is a waste of money and time. A gamble is reasonable if you can afford to lose without being ruined. The Supercollider was an unreasonable gamble because we could not afford to lose."

A second risk was that the machine had not been designed correctly to begin with. And in fact it had not been: after the cancellation of the project in 1993, the *New York Times* reported that the cost of its construction "jumped $2 billion in one fell swoop after designers discovered that the Supercollider's huge magnets . . . contained a hole too small for the proton beam to fit comfortably."

A $2 billion design error did not inspire confidence that the rest of the machine had been planned with any greater care or accuracy. Indeed, there was the additional design risk—the third in a long line—that one or more of the SSC's components would fail to work as advertised if and when the device ever became operational. Fermilab's rash of exploding magnets provided a clear precedent here. In the case of the SSC specifically, prior to its construction the Central Design Group had produced eight full-size examples of the machine's fifty-five-foot-long magnets. Only two of the devices worked, and they worked poorly at that. According to a report on the project, "The first was a slightly modified version of the planned production magnet, calling into question its relevance to the final product, and the second did not achieve its design strength the first time it was powered up."

A fourth risk was that even if every individual magnet did in fact work correctly, nevertheless two or more of them might fail to work in tandem when they were joined together and powered up. This, indeed, was exactly what would happen much later at CERN, in 2008, when a mere nine days into the operation of the Large Hadron Collider a helium leak damaged more than fifty superconducting magnets. According to the CERN press office, "the cause of the incident was a faulty electrical connection between two of the accelerator's magnets. This resulted in mechanical damage and release of helium from the magnet cold mass into the tunnel." The accident was followed by an operational delay of one year and two months before the machine once again resumed functioning.

Yet another risk, the fifth, was that, owing to its colossal size, the SSC would take so long to build (nine years or more) that during the interim it would be made obsolete by one or more accelerators elsewhere that would beat the machine to its goal. There was a precedent for this as well. In 1978 the Brookhaven National Laboratory had broken ground on a new proton-proton colliding-beam accelerator called Isabelle (Intersecting Storage Accelerator + "belle"). It was intended to discover the W and Z^0 bosons. In 1981, during the course of construction, production models of the machine's superconducting magnets failed even when energized at less than full strength. In 1983, when Isabelle was four years behind schedule, scientists at CERN managed to discover the W and Z^0 bosons, leaving Isabelle without a clear reason for being. In July of that same year, therefore, the High Energy Physics Advisory Panel (HEPAP), which in 1974 had recommended that Isabelle be built in the first place, now recommended that the device be canceled, which it was—"a waste of money and time."

But even that did not end the string of risks associated with the SSC. Another, the sixth, was that the results of the SSC's particle collisions might not be interpreted correctly. The SSC would collide hadrons, specifically protons, and proton-proton collisions were known to be especially tricky to decipher—much more so than electron-positron collisions, for example. The Large Hadron Collider at CERN, which was far smaller than the SSC was to be, had not one but two separate detector halls, and each was as large as a moderate-size office building: the CMS (Compact Muon

Solenoid), which was the size of a six-story building, and another named ATLAS, which was the size of a *nine*-story building. (Neither detector hall was very "compact." The ATLAS web page described the detecting apparatus as "about 45 meters long, more than 25 meters high, and weighs about 7,000 tons. It is about half as big as the Notre Dame Cathedral in Paris and weighs the same as the Eiffel Tower or a hundred 747 jets [empty].")

Nevertheless, enormous as they were, CERN's detector halls were pygmies compared to those planned for the SSC. According to an official construction manual for the project, "Contracting Practices for the Underground Construction of the Superconducting Super Collider" (1989), "Detectors for the SSC will far surpass existing collider detectors in both size and complexity. Resolving the precise paths of the hundreds of particles that can arise from a single proton collision and capturing the most energetic of these particles will require detectors that may exceed 40,000 tons."

Plainly, interpreting the SSC's results would not be a task for the timid, or for anyone in a hurry. As a case in point, the first traces of the Higgs particle were observed at CERN in July 2012, but discovery of the particle was not confirmed and formally announced until eight months later, in March 2013.

In view of this litany of hazards, which is by no means exhaustive, it is beyond doubt that building, testing, and operating the SSC, as well as interpreting the data stream provided by its proton-proton interactions, were all exceptionally risky ventures. But these difficulties were not often spoken of by the project's advocates, who tended to promote a success narrative in which the SSC's main problem was funding, after which everything else would fall into place and then sail along smoothly to the preordained grand conclusion: discovery of the Higgs particle and consequent vindication and refinement of the Standard Model. But such a rosy picture was hardly realistic. The SSC thus satisfied the third criterion of a pathological technology: the systematic minimization of its associated risks.

The fourth criterion is that a pathological technology's costs exceed its benefits by a wide margin. That the SSC was a costly project is manifest from the historical record. *The cost estimates of building the SSC never stopped rising and climbed inexorably in a series of growth spurts,* starting at a relatively cut-rate $3.9 billion, jumping to $4.9 billion, then $5.3 billion, $5.9 billion, $6.6 billion, $7.2 billion, $7.8 billion, $8.2 billion, and $9.2 billion, before topping out finally in 1991 at an estimated $11.8 billion. And that was well before the project was even one-quarter complete. Given that the global population in 1990, when the SSC was under construction, stood at approximately 5.5 billion, there was a short time period during which the cost of the SSC was equal to one US dollar for every living being on the planet. And by the time its cost had risen to $11 billion, that amount would have been equivalent to almost *two* US dollars for each man, woman, and child on earth. The question is whether the benefits of the machine would have been worth the money it would have taken to build it.

In 1988, before the machine had been authorized by Congress, the Congressional Budget Office (CBO) issued a report entitled *Risks and Benefits of Building the Superconducting Supercollider,* a document that ran to more than 100 pages. Because the CBO was by law nonpartisan, the report did not make any recommendations one way or the other. It nonetheless laid out a strong factual case for the conclusion that the costs of the machine would indeed have swamped whatever benefits it might have provided.

For one thing, "the SSC would consume a substantial portion of the current federal budget for basic science," and it was not obvious that that budget should be so radically skewed toward physics, since "it is difficult to say whether advancement in physics is more important for the advancement of human knowledge than advancement in biology or any other basic science." The fact was that high-energy physicists constituted a tiny fraction of the nation's scientists. "DOE estimates that currently 600 graduate students nationwide are studying high-energy physics. By contrast, in 1986, there was a total of 102,000 PhD students in science, excluding psychology and social sciences," the report stated. "In addition, the number of technicians working on high-energy physics projects is comparable to the number of graduate students studying high-energy physics. High-energy physics, therefore, does not provide a substantial number either of technicians or of graduate students, relative to its level of federal funding."

The CBO was not alone in making this argument. In 1987 the Nobel Prize–winning physicist Philip Anderson published "The Case Against the SSC" in *The Scientist*. He argued that "those who manage the funding of science have a very strong prejudice in favor of large expensive projects and large unitary laboratories or centers. The amount of money available for free, unprogrammed individual research dwindles. Unfortunately, it is not the large, expensive programs that produce really new things. . . . Innovative, small group work in this country must not be cut back in favor of large facilities."

Those who *were* in favor of large facilities often pointed to the unpredictable benefits that sometimes emerged from them, however serendipitously. Exhibit A in support of this view was the fact that the World Wide Web had been invented in 1989 by Tim Berners-Lee, who at the time was working at CERN. But precisely because they are fortuitous events, and thus unknowable beforehand, these benefits cannot be depended upon to happen, nor to be for the best if and when they do. Large projects could equally well have an entire range of unpredictable *negative* consequences: one technology that emerged from the American nuclear weapons program, after all, was none other than the nuclear geoengineering folly promoted by Project Plowshare. As for the World Wide Web, while today's civilization could hardly exist without it, it has also had its share of unintended negative consequences in the form of computer viruses, identity theft, and privacy invasions, as well as the massive hacking of major corporations, financial institutions, and governments.

And what of the main postulated benefit of the SSC, the discovery of the Higgs boson? The answer is that it did not take the SSC to discover it, the particle having been created and detected at the Large Hadron Collider at CERN. The SSC was not in fact needed to provide the principal benefit it was supposed to deliver.

Was the LHC itself a pathological technology? No, because it did not satisfy the four criteria; it certainly did not satisfy them nearly as well as did the SSC. The LHC was big, but it was less than one-third the size of the SSC, and it cost less than the SSC would have cost; in fact, it cost less than half as much, meaning that there was that much less at stake and less at risk.

In 1992, the year before Congress killed the SSC, Freeman Dyson said that "the Superconducting Supercollider oversteps the limits of reasonable size for a scientific project." Explaining his rationale, he said: "Even the scientists who believed that the Superconducting Supercollider was a splendid idea were aware that the big particle accelerators were running into the law of diminishing returns. Each step upward in the size of accelerators cost more money than earlier steps and took a longer time to produce new discoveries. The ratio of scientific output to financial input was diminishing rapidly as size increased. At some point not far in the future, the further growth of accelerators was bound to grind to a halt."

That was a prescient comment. In 2013, the same year in which the discovery of the Higgs was announced, twenty years after the cancellation of the SSC, and perhaps as few as *ten* years after the SSC might have been completed, accelerator technology witnessed one of the most stunning and remarkable developments in its entire history. *Suddenly, accelerators got smaller.* Not only smaller than the SSC and the LHC, but smaller even than Ernest Lawrence's original four-inch "proton merry-go-round." Accelerators had taken a quantum jump downward and were now *smaller than a grain of rice.*

Writing in the September 27, 2013, issue of the British science journal *Nature,* physicist Robert L. Byer of the Stanford National Accelerator Laboratory (SLAC) said, "The enormous size and cost of current state-of-the-art accelerators based on conventional radio-frequency technology has spawned great interest in the development of new acceleration concepts that are more compact and economical."

He and a team of experimentalists had in fact developed a new species of accelerator, one that sped up particles not by conventional radio waves but by pulses of visible light. By definition, light travels at the speed of light, and Byer and his group had utilized laser beams to accelerate particles to high speeds across small distances such as the interior of a silicon chip.

"We still have a number of challenges before this technology becomes practical for real-world use," said Joel England, a team member. "But eventually it could substantially reduce the size and cost of future high-energy particle colliders for exploring the world of fundamental particles and forces."

Nor was this a one-off flash in the pan. Simultaneously, a second group of researchers working at the Max Planck Institute of

Quantum Optics in Germany announced in *Physical Review Letters* that they too had succeeded in creating a proof-of-concept example of "an all-optical compact accelerator." Both teams envisioned a new generation of lab-sized, or even desktop-sized, particle accelerators at some point in the indefinite future.

And at that point, finally, particle physics might once again be small science, comparatively risk-free, and, not least, cheap.

Chapter 11

STARDATE 90305.55

According to the Star Trek Online Stardate Calculator, 90305.55 was the "stardate" on which the 100-Year Starship (100YSS) project began the first of its planned annual public symposiums, in Houston, Texas. (In conventional earthly terms, this was September 13, 2012.) The three-day conference brought together scientists, futurists, venture capitalists, college professors, students, attorneys, journalists, *Star Trek* fans, *Star Trek* stars, aerospace entrepreneurs, actors, actresses, entertainers, consultants, bloggers, science fiction writers, artists, ministers, members of the general public, and assorted others, virtually all of whom were intent on advancing a single ambitious, collective goal: defining a scheme by which human beings could leave planet Earth behind, travel to the stars, and establish a new Earth 2.0 in another solar system, all within the next 100 years.

The various means proposed to effect this dream were equally forward-looking. A NASA propulsion specialist described a mechanism for producing a space warp, the very same method used in *Star Trek* to flit through the galaxy in a jiffy. This "warp drive," he claimed, would allow for faster-than-light travel through the cosmos, exactly as in science fiction. A physicist whose doctorate was from Baylor University surveyed a range of alternative propulsion technologies, including fusion engines, antimatter propulsion, and pion (pi-meson) drives. His bold and radically "inclusive" approach was to make use of all possible physical fields from all possible physical dimensions, no matter whether they were known to exist or not. Still another scientist wanted to harness the somewhat abstruse and recherché process of Schwinger electron-positron pair-production to power a starship.

Despite the diverse backgrounds of the conference attendees, the madcap propulsion schemes of the physicists, and the literally outlandish nature of their common goal, it is important to keep in mind that the 100-Year Starship project was not created by or funded from the ranks of costumed *Star Trek* fandom, or by wild-eyed fanatics, amateurs, groupies, or lunatics, except perhaps in the strict Latin-derived etymological sense of *persons whose concerns pertain to the Moon*. As it happened, the 100YSS project had been conceived, implemented, and financed by DARPA, the Defense Advanced Research Projects Agency, a branch of the US Department of Defense, together with the NASA Ames Research Center, in Mountain View, California. Furthermore, the effort was headed up by Mae Jemison, MD, a former NASA astronaut, dancer, actress, foundation president, and educator, and the first black woman of any nationality to go into space.

The conjunction of all of these elements was slightly incongruous. How did a branch of the US military wind up bankrolling, to the tune of a half-million dollars, a group bent on going to the stars? Why were well-educated, mainstream, otherwise feet-on-the-ground, hard-nosed physicists and engineers describing starship propulsion systems that were so exotic that they depended on forces or factors that were not known to be real, utilized physical dimensions that were not known to exist, and were propelled by substances that had no plausible containment systems, existed only in nano amounts, persisted in time for only fleeting instants or minutes at best, and cost billions of dollars to produce?

Added to these anomalies was the fact that many of the conference presenters offered excellent reasons to believe that star voyagers would face a range of apparently insuperable and in any case potentially deadly threats en route: prolonged exposure to ionizing radiation, reduced life spans, boredom, alienation from the natural environment, fatal collisions with specks of interstellar matter, mass epidemics, the rise of a totalitarian, charismatic leader, crew mutiny, descent into crazed religious factionalism, and so on. Not to mention that age-old staple of science fiction movies: confrontations with a race of hostile, ugly, and all but omnipotent aliens bent on exterminating anyone who dared to invade their own private interstellar realm.

How, in the light of these and other obstacles, could audience members take the idea of star travel seriously? The answer was that they were under a spell, an enchantment; in their dream vision, we as a species were metaphysically fated to go to the stars. Indeed, what could possibly be more romantic, stirring, or compelling as a goal? Traveling to other star systems, it seemed, was our interstellar manifest destiny, our predetermined fate, a quest written in our DNA.

Neil Armstrong, the first man on the moon, once explained the allure of space travel in this way: "I think we're going to the Moon because it's in the nature of the human being to face challenges. It's by the nature of his deep inner soul. We're required to do these things just as salmon swim upstream." In other words, traveling to the moon was preordained, fixed, imposed upon us from without, not a venture of our own choosing. Like the functions of our autonomic nervous system—breathing, heartbeat, peristalsis—space travel was something that we couldn't refrain from, even if we wanted to.

And so, in the face of such a powerful, seductive, virtually irresistible psychological compulsion, was it at all surprising that physicists were proposing entire suites of pathological technologies to get us to the stars? We really had no choice in the matter. It was an imperative—obligatory, not optional. We as a species were headed for the stars, like it or not.

The idea of interstellar travel gained much of whatever credibility it had from the simple act of extrapolation from past experience. For example, there was the sequence of voyages from Columbus's transatlantic cruise to the New World, to Magellan's circumnavigation of the globe, to Lindbergh's solo flight from New York to Paris, to the Concorde (RIP), to Apollo 11, to the Pioneer and Voyager spacecraft, to . . . whatever came next in line. But what *could* come next other than traveling to Mars and the outer planets? And then, when we were finished with all that, heading off to the stars, which of course was our destiny as a species and written in our DNA. Not a matter of if but when. And so on through the latest wave of common platitudes.

Moreover, the human speed limit had jumped from walking and running, to the speed of a horse at full gallop, to 60 miles per hour on railroads and in early racing cars, to supersonic speeds in jet aircraft, to more than 25,000 miles per hour in rockets, to . . . whatever came next in line. But what *could* come next in line other than travel approaching, or even exceeding, the speed of light? After all, any number of popular science books insisted that travel at the speed of light was absolutely possible if you merely sized up the matter correctly.

Plus, there was the fact that people once naively believed all sorts of screwy and unscientific nonsense—that the earth was stationary and the center of the universe, that the earth's surface was flat, that anyone who went beyond the horizon would fall off the edge, and so on—but all of those quaint, charming, prescientific views had turned out to be wrong, so how was it possible for anyone today to believe that something would prevent us from traveling to the stars? After all, so much of current technology had once been regarded as moonshine—nuclear power, flight itself, human-powered flight, flying faster than the speed of sound, going to the moon, et cetera—that it seemed naive, backward, even churlish, and maybe even perverse and "anti-human" to hold a skeptical view of taking the next long leap, which was to the nearest convenient (actually, *inconvenient*) extrasolar luminous body.

Interstellar travel fans trotted out all of these time-honored, encrusted, venerable chestnuts, and many more besides, to buttress their pet notion that traveling to the stars was practically an inevitable development (and anyway, our destiny as a species).

Indeed, by the midtwentieth century, the idea of star flight was so well-entrenched and deep-seated a notion that the major milestones of the genre were themselves pretty much old hat. There was Freeman Dyson's Project Orion (circa 1968), according to which a starship would be propelled through space by a series of nuclear detonations acting on a pusher plate—a sort of pogo-stick rocket ship H-bombing itself through the cosmos. This was a device of which the mathematician Richard Courant once said, in his German accent, "Zis is not nuts, zis is supernuts." (A small non-nuclear prototype, which in flight tests had actually worked, was for a time on display at the Smithsonian's National Air and Space Museum in Washington, DC.)

There was the British Interplanetary Society's Project Daedalus (circa 1970s), a scheme in which a rocket powered by a nuclear fusion engine would travel to Barnard's star in fifty years or so. As a plan, this was admirable other than for the minor shortcoming that fusion engines did not exist at the time, nor do they yet.

And then there was the famed Bussard Interstellar Ramjet (circa 1960), an ostensibly elegant solution by which a starship would scoop up free hydrogen molecules from the interstellar medium and use them as a power source, supposedly gathering "fuel for free" as it sailed through space. As such, the device was a sort of cosmic perpetual motion machine. A later calculation by E. J. Öpik, however, found that in order for it to work as planned, the ramjet scoop (which Bussard envisioned as an electromagnetic field generated by the starship) would have to be more than 600,000 miles in diameter (more than twice the distance from the earth to the moon). A still later analysis by Thomas Heppenheimer ("On the Infeasibility of Interstellar Ramjets," 1978) showed that if the interstellar medium were "optically thick," the scoop would have to be half a light-year across, whereas if it were "optically thin," then the drag produced by a phenomenon called *bremsstrahlung* radiation would exceed the total amount of energy gained. More recently, writer Paul Gilster went so far as to say that Bussard's electromagnetic scoop "may turn out to have more applicability in braking than acceleration." So much for the perpetual-motion interstellar ramjet.

These three concepts, however, were merely the most prominent among the many competitors that vied for attention in the starship conceptualization sweepstakes. There were schemes for light-sail starships, laser-pushed light-sails, magnetic monopole rockets, resonant cavity thrusters (the "impossible" Cannae Drive)—far too many propulsion system ideas rather than too few. Indeed, beholding this avalanche of schemes was entertainment of a high order. As for developing the actual hardware to embody and enact any of this program, that was the furthest thing from anyone's mind.

Except the minds of DARPA planners and one or two other dreamers.

When DARPA announced the 100-Year Starship study in the fall of 2010, the agency's official press release gave rise to vast and mass confusion as to what it was actually proposing. The news release said: "The Defense Advanced Research Projects Agency (DARPA) and the NASA Ames Research Center have teamed together to take the first step in the next era of space exploration—a journey between the stars." From this it sounded as if the agency was proposing to plan and build, or at least underwrite the planning or building, of a starship. But DARPA was not in the spaceship business (nor was NASA Ames) and was also not in the least concerned with anything beyond the Earth other than military satellites. And in fact, constructing an interstellar spacecraft was not exactly the idea. In an attempt to clarify matters, the document stated further that "the 100-Year Starship study looks to develop a business case for an enduring organization designed to incentivize breakthrough technologies enabling future spaceflight."

Translated into English, DARPA's message here was that it and NASA had realized that nobody in their right mind formulated plans on anything like the 100-year time horizon that they thought would be required to build, outfit, and launch a manned interstellar vehicle. So they wanted to seed-fund some private group to do so, and for an essentially backdoor reason: namely, to reap whatever possible spin-off technologies might accrue from such an endeavor: "DARPA anticipates that the advancements achieved by such technologies will have substantial relevance to

Department of Defense (DoD) mission areas including propulsion, energy storage, biology/life support, computing, structures, navigation, and others."

In other words, DARPA was proposing to fund a private starship research group in order to gain what militarily useful technologies might be created as a by-product of the group's efforts to mount a trip to the stars. What a mess. But as a means of acquiring more sophisticated instruments of warfare, the idea was at least original.

That was in the fall of 2010. In the spring of 2011, DARPA announced a competition for an award of approximately $500,000, to be conferred upon whichever private entity made the best case for itself as a potential starship planner, proselytizer, builder, and launcher. The winner was former astronaut Mae Jemison, head of the Dorothy Jemison Foundation for Excellence. And by the fall of 2012, Jemison was fully in charge of a new, nongovernmental, nonprofit organization formally called (and trademarked as) the 100-Year Starship (100YSS). This, apparently, was the group that would take mankind's first giant steps on its journey out of the solar system.

The task facing Jemison's nascent organization was of no small magnitude. The core problem of interstellar flight was the fact that even the "closest" stars were incredibly far away. The nearest star, Proxima Centauri, was 4.22 light-years (39,900,000,000,000 kilometers, or 2,479,271,057 miles) from Earth. Even if a starship were to travel as fast as the Voyager I spacecraft—which has gone farther than any man-made object in history and which officially entered interstellar space in September 2013 as it continued to speed away from Earth at the rate of 38,698 miles per hour—it would not reach our closest extrasolar destination for more than 73,000 years.

Chemical rocketry, which, after all, was as old as ancient China, was hopeless for crossing such distances. At the Houston

100YSS 2012 Public Symposium, Richard Obousy, the Baylor physicist who was also the president of Icarus Interstellar, a rival band of space conquerors, explained that to travel to the "close" Centauri group of stars at even 10 percent of the speed of light, a rocket would require more chemical propellant than existed on planet Earth today. Hence the universal predilection among starship buffs for "exotic" propulsion systems.

The one big problem with exotic propulsion systems, however, was that they didn't exist as yet and might not ever; the only thing that actually existed were *ideas* for such systems and devices, each of which had its own considerable inventory of drawbacks. Antimatter propulsion, for example, a scheme whereby normal matter and antimatter would annihilate each other, thus producing enormous amounts of energy that could be used as thrust, faced any number of apparently insurmountable difficulties. One, since antimatter destroys conventional matter, it would be impossible to store it in normal physical containers. Two, for the same reason, it would be impossible to direct antimatter thrust by means of conventional, physical exhaust nozzles such as are commonly used in chemical rockets. Three, the amount of antimatter that has been created so far in particle accelerators is infinitesimal. In 2010 researchers at CERN managed to generate just 38 individual atoms of antihydrogen, and to do so they had to run the experiment 335 times.

"We're ecstatic," said Jeffrey Hangst, one of the researchers after that success. "This is five years of hard work." Even so, the CERN experimenters had managed to confine those precious 38 anti-atoms inside a magnetic trap for all of one-tenth of a second. (Later, CERN researchers trapped 309 atoms of antimatter for 1,000 seconds, i.e., 17 minutes, an accomplishment that was considered a very big deal.)

The dismal record presented by these anti-milestones somewhat dimmed the prospects of workable antimatter propulsion "systems" at any time in the near (or even far) future. Plainly, if interstellar flight was to have a prayer, it needed some kind of miracle device, a virtually unbeatable, absolutely phenomenal means by which an interstellar spacecraft could go from here to anywhere in no time flat. And of course, this being the realm of interstellar discourse, theory, and fantasy, there *was* in fact just such a device, or at least there was the *idea* of one: the Alcubierre warp drive.

In 1994, Miguel Alcubierre, a Mexican theoretical physicist studying for his doctorate at the University of Wales, Cardiff, published a paper titled "The Warp Drive: Hyper-fast Travel Within General Relativity," in the peer-reviewed technical journal *Classical and Quantum Gravity*. His approach was entirely novel: instead of trying to travel laboriously through interstellar space, he wanted to change the very contours of space-time, doing so in such a way that the gap between origin and destination would in effect shrink. "Create a local distortion of space-time that will produce an expansion behind the spaceship, and an opposite contraction ahead of it," Alcubierre wrote. "In this way, the spaceship will be pushed away from the Earth and pulled toward a distant star by spacetime itself. One can then invert the process to come back to Earth, taking an arbitrarily small time to complete the round trip."

By distorting the space-time around it, the spacecraft would arrive at its destination faster than light itself could travel in normal, unperturbed space. "The spaceship will then be able to travel much faster than the speed of light," he said.

Simple as that.

In interstellar-dreaming circles, this concept became known as the *Alcubierre drive*. For several years, it remained in the province of sheer theory and was subjected to analysis by both defenders and critics. In his original paper, Alcubierre himself conceded that his warp drive "has one important drawback," namely, that it required the existence of "exotic matter" in the form of negative energy. "However, even if one believes that exotic matter is forbidden classically, it is well known that quantum field theory permits the existence of regions with negative energy densities in some circumstances," he wrote.

Critics of the Alcubierre drive, on the other hand, made several potentially even more damaging points: that the amount of negative energy required to transport a small spaceship across the galaxy would be greater than the estimated total mass of the universe; that the crew members would not be able to send signals to the front of the distorted region of space-time, meaning that they would be unable to control, steer, or stop the ship; that extremely high temperatures inside the distorted region would melt the spacecraft; and so on.

Amid all of these conflicting claims and counterclaims, the Alcubierre drive remained essentially a mathematical fiction. But then, finally, a NASA physicist decided to settle the issue scientifically, which was to say, by means of experiment. This was Harold G. "Sonny" White.

Sonny White was the Advanced Propulsion Team leader of the NASA Engineering Directorate at the Johnson Space Center in Houston. Although he held a PhD in plasma physics from Rice University, he had a background in mechanical engineering and had worked as manager of the Space Shuttle's remote manipulator arm, a highly practical and relatively down-to-earth enterprise, at least as compared to star travel. But he had long been interested in interstellar spaceflight and was particularly intrigued by the Alcubierre drive. In 2006, White had an idea for reducing the concept to practice by creating an actual space warp in the laboratory. It would be only a tiny warp "bubble," less than half an inch in diameter, a volume in which space-time would be perturbed, or "warped," by as little as one part in ten million.

"It's very, very modest, a microscopic instance of this phenomenon, nothing that you would try and bolt to a spacecraft by any stretch," White said at the 100YSS 2012 Public Symposium. "It falls into the category of an existence proof. You either confirm or refute your physical interpretation of the mathematics. That's why we're trying to generate this little, tiny warp bubble."

The experimental apparatus was located in White's lab at the Johnson Space Center. The basic setup was straightforward enough.

"We have a ceramic capacitor ring that we charge up to many thousands of volts, to implement a potential energy that blueshifts that frame relative to where the capacitor is located," White said. "What the field equations predict is that the presence of that potential energy and boost will create a spherical perturbation. So that's what we're trying to measure, to see if we can generate that change in optical properties in that little spherical region."

The capacitor ring was like the wire loop that a child might swing through an arc to create soap bubbles. If, after the capacitor ring was energized, the two sides of the spherical space encapsulated by it were optically different, the disparity would be registered by a specially built warp field interferometer, a device that could detect and measure minute variations in electromagnetic wavelengths. That optical difference, supposedly, would be evidence of a tiny deformation of space (as well as a dream come true for several generations of *Star Trek* fans).

But even if White's experiment succeeded in establishing the requisite space-time distortion on the minimalist order of one part in ten million in a bubble half an inch across, the question remained whether such a phenomenon could be scaled up to the energy requirements of interstellar spacecraft that, as the astronomer Carl Sagan once said, were the size "of small worlds." Sonny White thought that even in the best-case scenario we would see something real based on this concept "maybe within my lifetime."

Despite having sounded that cautionary note, in June 2014 White released a set of images, produced by artist Mark Rademaker and graphic designer Mike Okuda, of a possible design for White's warp-drive spacecraft. The craft was called IXS *Enterprise* and had been based on Matthew Jeffries's 1965 sketches for the original *Star Trek* warp-drive spacecraft, which was also, and not coincidentally, named the *Enterprise*. In the realm of advanced technology, of course, pictures commonly antedated practice, as was shown, for example, by Leonardo's diagrams for flying machines. They were emotionally satisfying anyway, as they offered the illusion that these things are, or at least could be, real.

As for the warp drive, Alcubierre himself had long since given up on the concept. "It's a nice idea," he told a journalist for *Popular Science* in 2013. "I like it because I wrote it myself. But it has a series of limitations that I've seen through the years, and I don't see how to fix them."

As happens with pathological technologies, nowhere in this parade of essentially conjectural propulsion "systems," devices, and assorted notional gadgets was any attention given to their possible dollar costs. There were several reasons for this. For one thing, all of these technologies were, to date, strictly imaginary, so that any rational cost calculation seemed beside the point. It was like pricing out a rocket ship that was basically a cartoon. Second, there was the fact that talk about these speculative mechanisms was confined largely to the ranks of physicists and engineers, people who were concerned primarily, perhaps even exclusively, with the physical possibility of their special projects, not with how wildly expensive they might be. Finally, the allegedly ineluctable character of the goal ("our destiny as a species") made any consideration of cost essentially irrelevant. Since we were going to the stars anyway, period, no matter what, why even bother considering how much it would cost us to get there?

The whole notion that exploratory journeys such as star travel was somehow "written in our DNA" was never argued for, but merely asserted as if it were self-evidently true, which it was not. It was also rarely challenged. The fact was, however, that staying home and doing nothing was as deeply ingrained in human DNA as the practice of making extended voyages to far-flung destinations. For every Magellan, Marco Polo, or Captain Cook there were hundreds if not thousands of humans who never strayed far from their own place of birth.

In any case, a manned interstellar voyage would be a project of unprecedented scale and scope, one of the biggest efforts of any kind in human history, requiring a decades-long, sustained commitment of an unknown number of billions or trillions of dollars and an outlay of massive amounts of energy, manpower, and material resources that, if put together into a starship and launched, would be leaving the planet for good. In the face of this, DARPA wanted to have a stated rationale for this wholly implausible effort, whether for cosmetic reasons or as a means of getting the general public on board with the project, or both.

And so, starting on the evidently symbolic date of 1/11/11, DARPA and NASA Ames sponsored a "strategic planning workshop" at the Cavallo Point Lodge in Sausalito, California. Located at the base of the Golden Gate Bridge, of which it offered stunning views, the lodge was a posh resort where rooms started at $500 per night and where, indeed, anything seemed possible.

Here arrived some thirty "visionaries," including: four science fiction writers; astronaut Mae Jemison; extraterrestrial intelligence searcher Jill Tarter; Marc Millis, head of the Tau Zero Foundation (yet another group of interstellar travel enthusiasts); Peter Diamandis, X Prize founder; and the human genome sequencer and, more recently, universal adviser Craig Venter to address, among other puzzles, the vital question: Why go? Typically for a pathological technology, none of the reasons offered for developing an interstellar vehicle actually withstood analysis.

For example, the top answer by far was that we would go to the stars *to ensure human survival and the future evolution of our species.* The reasoning here was that the earth might be made uninhabitable by any number of catastrophic events: by meteorite strikes, global warming, lethal disease pandemics, thermonuclear war, extraterrestrial invasion—all the mainstay entries in the contemporary lexicon of disasterism. Taking humanity to another star system would avoid all such planetary wipeouts.

As a motivation for interstellar travel, however, such a justification was decisively overturned by the simple fact that this same goal could be accomplished by placing human settlements *anywhere else in our own solar system:* on the moon, on Mars, in O'Neill-type space colonies (all the rage in the 1970s) scattered out among the asteroids, on the satellite Europa, et cetera. Moving humanity off the earth did not even remotely require "going to the stars"; it simply meant *going elsewhere.*

This was an important, indeed an absolutely crucial, point. Escaping an earthly catastrophe merely required leaving the planet. Where to end up was then an open question. All of the space surrounding our own solar system, and all the astronomical bodies within it, or that could be constructed within it, were possible destinations, even though some of them provided more favorable environments than others. Traveling to the stars was by no means necessary.

A second reason advanced for star travel was to further the process of human evolution, but that too could happen as easily on a planet (or an artificial habitat) orbiting the sun as on one revolving around another star elsewhere in the galaxy. Moreover, there was good evidence that humanity was constantly evolving, and always had been, right here on earth itself, where it could continue to do so, particularly in the age of synthetic biology.

A third reason the visionaries offered for star travel was to contact new life forms. To this rationale there were a number of answers, the first being that simple *communication* could accomplish that goal, rendering interstellar travel wholly unnecessary. Moreover, there were *risks* in contacting new life forms, particularly if the aliens were unfriendly, murderous, and virtually omnipotent masters of their own stellar empire, a scenario that had been depicted ad nauseam by any number of science fiction writers over the decades. Finally, there was the fact that there was as yet absolutely zero empirical evidence that other life forms of any sort, from microbes to monsters to advanced, super-intelligent beings, or smart computers, actually existed anywhere else in the universe. As a reason for interstellar travel, then, this one was a nonstarter.

One of the participants gave as his own personal reason for going to the stars: "Because it's just so damn cool." But coolth, like beauty, lay in the eye of the beholder. Alternative views of star travel could be, with equivalent logic, or lack thereof: "It's just so damn . . . expensive . . . unnecessary . . . unrealistic . . . stupid."

Pathological technologies are typically put forward, promoted, and developed despite the presence of substantial drawbacks or risks that, when considered at all, are commonly dismissed, downplayed, or passed over in silence by proponents. In the case of interstellar travel, the possible payoffs were substantially offset by a roster of downsides, dangers, and existential threats.

One category of risk arose from factors or forces external to the spacecraft—for example, collisions with objects in the interstellar medium. Collisions in space are by no means rare: by the end of the Space Shuttle program, for example, more than 100 shuttle windows had been replaced after impacts with space debris, some objects being as small as the fleck of paint that cracked the front window of STS-7 (the second *Challenger* mission) in 1983. After a while there had been such a rash of debris impacts that the shuttle, once it reached orbit, was intentionally flown tail-first to minimize the effect of collisions.

It might be thought that the interstellar medium is "empty space," or a vacuum. To the contrary, the space between the stars contains volumes of interstellar gas and dust, along with cosmic rays, and possibly objects of unknown composition, size, mass, and density. And so it would be difficult to believe that on a journey of at least 4.22 light-years (the distance from Earth to Proxima Centauri) an interstellar spacecraft would meet with no other object whatsoever. But for a starship traveling at relativistic speeds, a collision with even a random small particle, according to Tom W. Gingell of Science Applications International Corporation, who did a study of the subject, would have the effect upon the spacecraft of an H-bomb explosion.

Since quickly diverting a massive spacecraft from its course would be impossible, it would be necessary instead to detect, deflect, or destroy the object within a matter of milliseconds before impact, by means of a system that would have to work perfectly and virtually instantaneously the first time out. But no such highly sensitive, fail-safe, and fast-acting detection and deflection systems existed or were in prospect.

A second category of risk consisted of threats arising from within the spacecraft itself to the physical and mental well-being of those aboard it throughout the whole of its interstellar journey. Since the size, structure, internal environment, and population of the starship were unknowns, any attempt at assessing what the probable onboard living conditions would be like was essentially an exercise in guesswork. Although there were numerous designs of interstellar craft on paper, there was one concept that had remained relatively consistent across time: that of a space ark. An early description of it was given by the British crystallographer J. D. Bernal in his 1929 book *The World, the Flesh, and the Devil,* a volume that Arthur C. Clarke once described as "the most brilliant attempt at scientific prediction ever made."

In it, Bernal asked the reader to "imagine a spherical shell ten miles or so in diameter, made of the lightest materials and mostly hollow. . . . The great bulk of the structure would be made out of the substance of one or more smaller asteroids, rings of Saturn or other planetary detritus. . . . The globe would fulfill all the functions by which our earth manages to support life." The spherical shell, he added, would have "twenty or thirty thousand inhabitants."

As ahead of its time as it was, this was a concept with legs, because more than eighty years later, in *Scientific American*'s January 2013 "Future of Science" issue, anthropologist Cameron Smith wrote that "a Space Ark, a giant craft carrying thousands of space colonists on a one-way, multigenerational voyage far from Earth," is "technologically inevitable."

Neither Cameron Smith nor J. D. Bernal gave any reason to think that a space ark (also known as a Bernal sphere) was even technologically *possible,* much less "inevitable." But Bernal himself acknowledged that life aboard "would be extremely dull, and that the diversity of scene, of animals and plants and historical associations which exist even in the smallest and most isolated country on earth would be lacking." Life inside the starship would also be physically onerous, given the fact that these interstellar travelers would be confined to an artificial environment for their entire lives and would be subjected to an array of hazards, including cosmic ray damage and other forms of ionizing radiation, DNA mutations and cellular injuries, surface outgassing, epidemic diseases, possible mechanical failures (including the temporary or permanent breakdown of atmospheric, water, agriculture, waste recycling, or filtration systems), computer malfunctions and software glitches, crowding, lack of privacy, protracted isolation, boredom and unbearable tedium, trance states and depression, unforeseen emergencies, and every other source of misery and conflict that is found on earth (not excluding crimes of violence, murder, and suicide), although now with the further added attraction that the travelers would be locked up inside a vehicle ten miles wide, from which there was no escape, a craft that would be hurtling through a black void for years or decades, while out of real-time communication with the home planet and beyond any realistic possibility of rescue or outside assistance.

In addition to those physical, medical, psychological, and emotional dangers there would be ample opportunity among passengers and crew members for the rise of charismatic fundamentalism and/or other forms of religious fanaticism. After all, where human beings went, so too did their religions, even in the space age, as the history of American spaceflight has well demonstrated. During the Apollo 8 mission in 1968, and while in orbit around the moon, astronauts Bill Anders, Jim Lovell, and Frank Borman took turns reading verses 1 through 10 of the Book of Genesis. Later, after the landing of Apollo 11, Buzz Aldrin, the second man on the moon, administered the Eucharist to himself during his short stay on the surface. As he later revealed to a Christian periodical: "It was interesting for me to think: the very first liquid ever poured on the moon, and the very first food eaten there, were the communion elements." So to imagine that religion would disappear aboard a starship was as much science fiction as all the rest of the improbable interstellar scenario.

Further, in order for a space ark mission to be successful, there would have to be some sort of onboard government, including a constitution and legal system, a police force, courts, judges, juries, and prisons, both for purposes of effective law enforcement and to avoid tyranny, mutiny, civil war, or other forms of chaos, unrest, or revolt. As human beings, we would be bringing all of our conflicts, differences, and sources of division along with us, albeit now while confined inside an inescapable pressure-cooker located somewhere in space. And as was well known from all earthly experience, governments and their various subsidiary elements would be subject to corruption, rebellion, replacement, and destruction.

Still, in the unlikely event that the starship made it all the way through interstellar space without hitting anything, without the crew members suffering mass sickness, hallucinations, or epidemic death, civil war, or descent into religious fanaticism or ordinary secular madness, what an embarrassment it would be for the voyagers to discover, upon reaching their coveted new home in space, Earth 2.0, that unfortunately it was already occupied by a race of intelligent aliens who were all too ready, willing, and able to protect their turf by blasting these intrusive space invaders to smithereens or, worse, vaporizing them into sheer nothingness.

One might suppose that this possibility would have been entertained and guarded against well before launch, and that the travelers would have verified that their prospective extrasolar earth was uninhabited, or at least that any natives in the area were friendly, warm, and charming, with a wide variety of interesting restaurants. But any number of difficulties would suffice to throw cold water upon that rosy picture. For one thing, since the extrasolar planet was an unknown number of light-years distant, there would be a substantial delay between the possible sending of an unmanned exploratory probe and the arrival back at home of its report, during which time conditions on the target planet could have changed to the point that it could have been colonized by aliens who had already exhausted its resources, turned it into a trash dump or a penal colony, moved it out of its orbit, or destroyed it altogether.

Other scenarios were, of course, possible. An exploratory probe could in fact fail to discover hostile natives because they were intentionally hiding themselves from detection or even posing as friendly in order to lure starry-eyed intruders to their doom. There was in fact *no* fail-safe way of knowing what the conditions would be like on the target planet before the starship's actual arrival there, at which point it might be far too late for the space invaders from earth to go home again, or to go elsewhere.

Finally, supposing that none of these catastrophes ever happened and that our group of interstellar voyagers, or more likely their remote descendants, reached their destination safely and intact; what would they then *do* with the pristine and virgin extrasolar planet they colonized? Assuming that human nature had not changed in the interim, they would in all likelihood do with it more or less what the earth's native population had long since done to their own home planet, which was a matter of common knowledge and considerable embarrassment to many.

But there was another, closer analogue to the probable legacy of human action on a another celestial body, to wit, our treatment of our own celestial companion, the moon. By the end of 2012, human beings had deposited approximately 400,000 pounds of

man-made objects, debris, and space junk on the moon. Most
of the total was spacecraft wreckage, the bits and pieces of more
than 70 rockets that had landed upon or crashed into the moon,
starting with the 800-pound Russian Luna 2 vehicle in 1959. In
addition, there was the collection of objects left behind by the
Apollo astronauts, who took more than pictures and left a lot
more than footprints. Indeed, more than 100 items were abandoned or intentionally placed on the Sea of Tranquility by Neil
Armstrong and Buzz Aldrin alone, including shovels, rakes, an
American flag, geological tools, lunar experiment packages, re-
flectors, and the Lunar Plaque that read: HERE MEN FROM THE
PLANET EARTH FIRST SET FOOT UPON THE MOON JULY 1969, A.D.
WE CAME IN PEACE FOR ALL MANKIND.

But did we leave the place as good as, or even better than,
we found it? Was the moon actually *improved* by our presence?
Waves of later American astronauts left on the moon's surface a
succession of lunar landing modules and rovers, tongs, scoops,
golf balls, twelve pairs of boots, television and still cameras,
lenses, film magazines, lines and tethers, boxes, canisters, cables,
filters, antennas, packing material, hammers, cockpit seat arm-
rests, ninety-six bags of human urine, feces, and vomit, blankets,
towels, wet wipes, personal hygiene kits, empty packages of space
food, and various sculptures, pins, patches, medals, and miscellaneous other pieces of flotsam, jetsam, and junk.

This, then, was the situation. There appeared to be no good
reason for going to the stars to begin with, and no good way
to get there, particularly in any sort of reasonable, human time
frame. A starship would confront ample physical dangers from
without, while its population would face substantial hazards
of their own arising from conditions inside the ship itself. The

The lunar landfill: the moon's surface as left by the crew of Apollo 17, December 1972.

interstellar travelers could face further unpleasant surprises upon arrival at their destination.

If the trip were a multigenerational journey, then no one who departed on it would be alive upon the ship's arrival at its destination, meaning that nobody who boarded the craft would benefit from the trip. Nor would anyone on earth who witnessed the launch, or helped pay for the craft, be alive to behold the ship's grand appearance at the distant star system.

What, then, was the point of it all? In the annals of pathological technologies, there was no other idea that depended for its plausibility on such an extreme degree of juvenile aspiration, magical thinking, systematic denial, and sheer silliness than the imaginary technologies of interstellar flight.

It did not follow from this that manned spaceflight per se was inherently pathological, nor that any or all of the exploits in its past history fell into that category. The Apollo program, for example, did not meet the criteria of pathology, for two main reasons. Although it was wildly expensive, the cost was not out of proportion to its benefits, which included, among other things, clear proof that human travel to other celestial bodies was in fact possible. If humans were to migrate from the home planet, to Mars or elsewhere, then the moon flight was the first and most important milestone in that direction. Second, the risks of manned spaceflight were too obvious, substantial, and well known for them to be ignored, minimized, or papered over. Indeed, in the public addresses in which he made a case for going to the moon, President John F. Kennedy not only portrayed the moon flight as a stepping-stone to the rest of space but also acknowledged its difficulty. In his first "moon speech," delivered to a joint session of Congress on May 25, 1961, Kennedy famously said: "I believe this nation should commit itself to achieving the goal, before this decade is out, of landing a man on the moon and returning him safely to the earth. No single project in this period will be more impressive to mankind, or more important for the long-range exploration of space; and none will be so difficult or expensive to accomplish."

In the second such speech, at Rice University on September 12, 1962, Kennedy emphasized the difficulty of pioneering space travel: "We choose to go to the moon in this decade and do the other [difficult] things, not because they are easy, but because they are hard." They were hard not only because they were scientifically and technically challenging, but also because manned spaceflights were risky ventures. Astronauts or others involved in the enterprise could die horrible deaths (which some of them did), the entire project could fail, rockets could get lost in space. Pictures of spacecraft exploding on launch pads were pervasive in the mass media, and the threats they posed to astronauts were impossible, even for politicians, to cover up or deny. Anyone who has read *The Right Stuff,* Tom Wolfe's legendary history of the Mercury program, will recognize the oft-repeated refrain, "Our rockets always blow up," which, as a statement of fact, was only slightly hyperbolic.

The situation has not changed significantly during these early days of *private* spaceflight, a set of ventures that are neither large in scale nor all that expensive: the entire private launch industry is minuscule, far smaller in comparison to NASA-sponsored space-flight than is private aviation as compared to commercial air travel.

Nor are the risks of private trips into space hidden from the public, ignored, or downplayed. Indeed, it would be impossible to do so: the fatal crash of Virgin Galactic prototype vehicle Space-Ship Two, the VSS *Enterprise,* in October 2014 was so widely publicized in all forms of media that any attempt to soft-pedal the dangers of commercial space flights would be ludicrous. For these reasons, "space tourism" is not even remotely pathological. Tech-nologies may, of course, be criticized on other grounds than being pathological: they might be pointless, naive, poorly executed, silly, and many other things, but none of these faults need make them pathological enterprises.

The Space Shuttle and the International Space Station (ISS) are harder to classify definitively. Both have been big, expensive programs, and the risks of shuttle flights were minimized until the *Challenger* and *Columbia* disasters made that procedure impossi-ble. There is also the problem that the scientific returns of both projects are disproportionately small in relation to their expense, and the likelihood of a big payoff is remote. The two projects seem to be mired in a loop of mutual justification: the shuttle exists to supply the ISS, whereas the ISS exists to justify the shuttle's re-supply missions, the whole circular operation existing mainly for the purpose of maintaining itself indefinitely as a going concern. (The principal scientific justification for the space station was to study the effects of long-term exposure to zero-gravity conditions on the human body.)

Traveling to the stars, by contrast, suffers from no such am-biguities or uncertainty. Indeed, it is a special case of manned spaceflight, more daunting than near-Earth spaceflight by many orders of magnitude. Star travel in fact occupies a special niche in the long career of human aspiration and desire. But although one of the most commonly expressed motivations for "going to the stars" is to perpetuate the human race, it is far more likely that an interstellar voyage would mean not the survival but rather the death of its crew. Arguably, the building and launching of a manned interstellar starship would be one of the most waste-ful, expensive, dangerous, and foolish projects in all of human history—a pathological technology for the ages.

PART III

ENDINGS

Chapter 12

THE SIX *HINDENBURGS*

The *Hindenburg* disaster occurred on May 6, 1937, which was a Thursday. Almost immediately the US Department of Commerce appointed an official commission to investigate the tragedy, and its members gathered together for the first time just four days after the incident, on Monday, May 10. Separately, the Federal Bureau of Investigation started its own in-depth inquiry; the agency would end up compiling a voluminous classified report on the case. Despite these official investigations, the conclusions drawn, and the evidence made public, the *Hindenburg* disaster soon became a pop culture "mystery," an item on a par with latter-day "mysterious" events or phenomena such as the Bermuda Triangle, Area 51, and the UFO incident in Roswell, New Mexico.

One reason for this aura of mystery was the fact that the *Hindenburg* fire was one of the first major disasters to be captured on film, "live, as it happened." The power and immediacy of that recorded visual display seared the event into popular consciousness and collective memory. Allied to this was the fact that all the known depictions began just *after* the fire had started, meaning that no newsreel footage or still images showed the moment of ignition, the first appearance of an explosion, or the beginning of a flame front. Indeed, there seemed to be two separate visual histories that did not match up or connect with each other in any way. The first was that of an intact airship gliding silently toward its mooring mast with all the assurance of an ocean liner. The second was that of an airship ablaze, cracked in half, and falling out of the sky. It was as if the *Hindenburg* had entered a solid wall of space and time and then burst out through the other side, with everything that happened in between permanently hidden behind some sort of impenetrable event horizon.

The unfilled space that separated these two trains of events was deeply unsatisfying and gave rise to the *Hindenburg* "mystery," which persists to this day.

Back in 1937, however, the Department of Commerce hoped to solve it. The investigating commission met for the first time on Monday morning in a portion of the Lakehurst hangar that had been used as a waiting room for zeppelin passengers and after the disaster was converted into a temporary morgue. The commission members included several Commerce Department officials, aided and abetted by technical advisers from the army and navy, as well as civilian specialists in aviation and aeronautics. Also a member of the American investigating team was Lakehurst commander Charles E. Rosendahl, who had beheld the calamity from beginning to end and also participated in it, having recommended that Max Pruss make the "earliest possible landing" despite marginal weather conditions.

Other actors in the drama—surviving passengers and crew members, including Helmut Lau, Hans Freund, and Heinrich Kubis, among others—would also give evidence, as did members of the ground crew and even a few spectators. In addition, a parade of expert witnesses testified regarding the weather, electrostatic phenomena, and the workings of explosive devices.

On the fourth day of testimony—slightly more than a week after the accident—the six-member German committee arrived at Lakehurst. The principal members of the group were Hugo Eckener, director of the Zeppelin Company; Ludwig Dürr, chief designer of the *Hindenburg*; and Max Dieckmann, a professor and an expert in atmospheric static electricity.

Over the next three weeks this international team of experts considered virtually every hypothesis, no matter how unlikely, that could explain the *Hindenburg* fire. A year later, in August 1938, the Department of Commerce published *Report No. 11: The* Hindenburg *Accident,* consisting of a translation of the twenty-one-page German report and the fuller, sixty-eight-page American account. The two documents agreed in almost every particular, including the final conclusion of both investigative teams that the proximate cause of the accident remained unknown. Likewise, both groups claimed that, owing to lack of evidence, sabotage was improbable, although in view of the fact that no definite cause could be established with certitude, it could not be ruled out as a possibility. Nevertheless, the German and American commissions each furnished a most probable cause, and in this, surprisingly enough, they differed.

The Germans rejected sabotage as the cause, whether perpetrated from inside the airship, as from a timed explosive device, or from outside, as in the form of incendiary bullets fired from the ground or, what was even more unlikely, from an aircraft. On the basis of experiments that had been performed in Germany, they eliminated sparks from the diesel engines or from their exhaust gases as possible sources of ignition. The Germans also considered and discarded the possibility of radio waves, propeller breakup and penetration of the hull, a spark from the breaking of a wire internal to the ship, and even the extremely remote chance of hydrogen gas self-combustion.

The Americans, in addition to rejecting these possibilities, also dismissed the prospect of a "high powered electric ray," as well as major structural failure, a spark from the ship's electrical system, and ball lightning.

The two teams agreed that hydrogen had been leaking from one or more cells near the area where the craft's vertical fin intersected the top of the hull. That was where witness R. H. Ward, a ground crew member who was in charge of the port bow landing party, had seen a fluttering of the ship's outer skin just a few seconds before the fire. "In his opinion the motion of the surface was not due to the slipstream or resonance effect of the propeller," the American report stated. "It was entirely too high from the propeller. It appeared to him to be more like an action of gas inside pushing up, as if gas was escaping. . . . The flutter was followed by a ball of flame approximately 10 feet or so in diameter; then came an explosion."

Further, "Eckener said that a leak in a gas cell, permitting the escape of 40 to 50 cubic meters of gas per second, would be sufficient to cause a flutter in the outer cover which could be observed as reported."*

---

*The phrase "cubic meters of gas per second" is evidently a mistranslation or a transcription error. Gas exiting the ship at such a rate would be truly catastrophic and would have exhausted all seven million cubic feet of hydrogen (200,000 cubic meters) in a little more than an hour, a release that would have been registered instantly on the control car's gas board, which it was not. A leak rate of forty to fifty cubic meters *per minute* was most probably what Eckener had in mind.

The escape of hydrogen gas into free air would have been necessary for the fire to have occurred. Hydrogen required the presence of oxygen in order to burn, and for this reason the Germans had always prided themselves on filling a zeppelin's gas cells with the purest hydrogen available.

The *Hindenburg* "mystery" thus reduced to the question of what had ignited the oxygen-hydrogen mixture. For both the Germans and the Americans, the answer was: some sort of electrostatic phenomenon. Prior to the fire, the atmosphere surrounding the *Hindenburg* had been brimming with electrical activity. After its first approach to Lakehurst, the *Hindenburg* had left the area and gone on a grand tour of the New Jersey coastline precisely because of thunderstorms over and around the landing field. Indeed, when Rosendahl radioed to Pruss that conditions were "suitable for landing," he had added the proviso:

THUNDERSTORM OVER STATION

And even as the ship finally made its approach to the mooring mast, the lightning of a secondary thunderstorm was visible off in the distance. The atmosphere immediately around the ship, then, had been highly electrically charged—and so too, therefore, was the *Hindenburg* itself.

While both the atmosphere and the ship had been positively charged, the earth itself was negatively charged. The opposed charges would have been equalized when the ship's landing ropes fell to the earth, electrically grounding the craft. Under those conditions, the American report said, "a spark might have been created."

But in a final incongruous twist, the Americans ended up attributing the ignition of the oxygen-hydrogen mixture *not* to a spark caused by the grounding of the ship, but rather to St. Elmo's Fire, a discharge of static electricity from objects in or near a thunderstorm, even though the American investigators conceded that "no witness testified that a visible indication of it was present."

The Germans, however, would have nothing to do with St. Elmo's Fire. For them, the most probable cause of the *Hindenburg* disaster was the electrical grounding of the ship, which "could lead to equalization of tension by a spark, which possibly caused ignition of a hydrogen-air mixture present over the gas cells 4 or 5."

There was one apparent inconsistency in the Department of Commerce report, and that was the precise location of the original ignition point. The four men in the lower fin—Helmut Lau, Hans Freund, Rudolf Sauter, and Richard Kollmer—all testified that the fire had started practically right before their eyes, which was to say, at or near the centerline of the ship. But eyewitness accounts from outside the vessel—those on the ground, including Commander Rosendahl, as well as the civilian spectators standing beyond the naval base fence—all reported seeing a flame at the very *top* of the ship, at or near the point where the upper vertical fin entered the ship's backbone.

According to the Americans, there was no need to reconcile these apparently conflicting accounts, because there was in fact no conflict: "The first open flame, produced by the burning of the ship's hydrogen, appeared on the top of the ship forward of the entering edge of the vertical fin over Cells 4 and 5," which was to say, where Rosendahl and the others saw it. "The first open flame that was seen at that place was followed after a very brief interval by a burst of flaming hydrogen between the equator and the top of the ship," where Lau, Freund, Sauter, and Kollmer saw it. "The fire spread in all directions, moving progressively forward at high velocity with a succession of mild explosions." In other words, the two versions of events were in fact two parts of the same phenomenon, separated slightly in time and observed from two different perspectives.

And with that, the explanation of the *Hindenburg* disaster seemed to be complete. The German and American reports jointly accounted for all of the observed phenomena and contradicted none of them, and they were consistent with all the known and established facts of the case.

Still, there were two matters left dangling. One was the puzzle concerning the two possible causes: was it static electrical discharge as a result of grounding, or was it St. Elmo's Fire? (It was theoretically possible that both phenomena had occurred simultaneously, and that both had played a role in starting the fire.)

The second was the fact that sabotage could not be ruled out entirely under any interpretation of the events in question. Thus the "mystery."

To many old-line zeppelin die-hards, captains, and other crew members, there *was* no mystery. These people found it difficult to accept that one of their cosmic airships could be defeated by mere forces of nature such as lightning, atmospheric electricity, St. Elmo's Fire, or any other such thing. "The Zeppelin family believed completely in the soundness of the *Hindenburg* and in their ability to operate it safely with hydrogen," said airship historian J. G. Vaeth. "If something went wrong, it must have been the work of outside agents. This mindset led many, Pruss included, to insist that the ship had to have been sabotaged."

It was not only Max Pruss who insisted on this; so did former *Hindenburg* captain Ernst Lehmann, chief purser Heinrich Kubis, and Charles Rosendahl. Kubis was even convinced that he knew the identity of the saboteur: the acrobat Joseph Späh.

After all, it was Späh who, during the voyage, had made anti-Nazi jokes and wisecracks and kept traveling to the rear of the *Hindenburg* to visit his performing dog, Ulla, even in the face of repeated warnings from Kubis not to do so unaccompanied. On top of that, Späh was an accomplished gymnast who could easily climb ladders, as well as crawl all over the ship's metal framework, the better to plant a bomb. There were problems with this theory, however, the first being that there was no evidence that Späh had in fact possessed such a device, much less set one in place, nor that one had ever existed aboard the *Hindenburg*. Nor was Späh known to have a motive.

For all of its supporters from within the zeppelin ranks, the sabotage theory was never taken very seriously by others until the publication of the book *Who Destroyed the Hindenburg?* in 1962, twenty-five years after the event. The book had been written by A. A. Hoehling, the author of several previous works of popular history. In alleging that the ship had been intentionally destroyed, Hoehling claimed that, prior to the *Hindenburg* fire, the Germans themselves had regarded sabotage as a distinct possibility. The German embassy in Washington, he said, had received "hundreds" of anti-Nazi letters threatening the *Hindenburg*'s destruction, and in fact the author prefaced his book with one of them. Written by Kathie Rauch, of Milwaukee, Wisconsin, and dated April 8, 1937 (a month before the disaster), the letter said in part that "the Zeppelin is going to be destroyed by a time bomb during its flight to another country." Supposedly, Captain Lehmann had carried a copy of this letter on his person when he boarded the craft in Frankfurt. And it was precisely to deal with the bomb threat that six captains of the line were aboard the *Hindenburg* on its final flight.

The Germans had taken these threats so seriously that customs officials had searched the *Hindenburg* itself, its passengers, and their luggage for contraband—including, presumably, bombs—prior to the start of the final flight. Still, in view of the vast size of the craft, any search of the ship was bound to be incomplete. Hoehling claimed that on one of the *Graf Zeppelin*'s trips back from Rio, five monkeys had been smuggled aboard the craft. And contrary to strict Zeppelin Company policy, cigarette butts had also been seen on catwalks below the *Graf*'s gas cells, as if crew members had smoked routinely. It would be easy, therefore, for a disgruntled or deranged crew member to sneak a time bomb aboard the *Hindenburg*.

But the shock was the identity of the person Hoehling had accused of doing these things: Erich Spehl, the young rigger who had been at the very bow of the ship prior to its destruction, who had been horrifically burned by the column of flame that had issued from the nose, and who had died of his burns during the night. The only evidence that Hoehling produced against Spehl were the claims that he was "moody and possibly introverted," that he had a Communist girlfriend who wanted to embarrass the Nazis by destroying the *Hindenburg*, and that he was an amateur photographer who could have constructed a detonating device with a flashbulb, battery, and timer. As true as these claims might have been, there was nevertheless no proof that Spehl or anyone else had actually constructed or planted such a device.

Finally, the warning letter that had been mailed to the German embassy in Washington was real enough, but as Hoehling himself acknowledged, the scenario it depicted had originated in a dream and its sender, Kathie Rauch, was a known spiritualist. With elements of the supernatural now coupled to a lack of factual evidence that the *Hindenburg* had been sabotaged, much less that Erich Spehl was the perpetrator, this was a story ripe for Hollywood.

Ten years after Hoehling's book, author Michael Mooney published a highly fictionalized follow-up account, entitled simply *The Hindenburg*. Mooney's apparent knowledge of Erich Spehl and his movements aboard the craft bordered on the omniscient: "Shortly before going off watch, Spehl cut the fabric of Gas Cell IV, deep down within the folds of the loose bag, and started the timer. He set it at 2 hours—that would be eight o'clock. He rearranged the folds of the drooping cell to hide the slice. Since the hydrogen was pressing up to the top of the cell, hardly any would leak at the bottom where he had made his cut—until the phosphorus burnt away the fabric and the air began to pour in. Then the oxygen and hydrogen would meet and—*poof!*"

Poof! indeed. While Mooney's book was still in outline stage, its publisher, Dodd, Mead & Company, entered into an agreement with Universal City Studios according to which Universal would acquire motion picture rights to the book, would promote it, and would pay Mooney a royalty based on sales. Which is how it happened that the film version, also called *The Hindenburg*, with an all-star cast featuring George C. Scott, Anne Bancroft, Burgess Meredith, and Gig Young, was released in 1975. Erich Spehl, who in the movie was renamed Karl Boerth, was played by William Atherton. George C. Scott portrayed a fictional Nazi intelligence officer working for Propaganda Minister Joseph Goebbels. Scott is shown as having located the time bomb at the last minute, but failing to disarm it, he gets blown to bits.

With two books and a movie going for it, sabotage became the orthodox theory as to what had destroyed the *Hindenburg*. Ten years later, though, that theory had been replaced by a far more unlikely explanation. According to its originator, a retired NASA scientist by the name of Addison Bain, "The presence of hydrogen in the lift cells was not the starting point of the fire of LZ 129, and it did not contribute too much to the further course of the accident. . . . The role of hydrogen was negligible. That hydrogen was the culprit is probably a pretext, in any case a myth."

Most startling of all, Bain denied that hydrogen combustion was to blame for any of the *Hindenburg*'s thirty-six deaths. "The fire of hydrogen from the gas cells lasted only less than one minute," he said, "and there is no evidence that anybody was directly hurt by it."

Truly, this was an amazing theory. But if hydrogen did not cause the *Hindenburg* disaster, nor even "directly hurt" anyone, then what did? Addison Bain's answer was, of all things, paint. "The paint of the outer cover was a highly flammable mixture," he said. "The first layer contained iron oxide. Then followed five layers of cellulose butyrate acetate with aluminum powder added." This meant, he claimed, that the mixture was a chemically close relative of the fuel used in the Space Shuttle's solid rocket boosters. "LZ 129 was literally painted with rocket fuel."

Everyone has heard of Holocaust deniers and global warming deniers. Here, apparently, was a hydrogen denier.

In actual fact, Addison Bain was a longtime *proponent* of hydrogen as an alternative energy source and was a founding father of the National Hydrogen Association. His anti-hydrogen theory was a big hit with others of like persuasion, and in October 1998, a year after Bain's revisionist account had been featured in an issue of *Popular Science* magazine under the title "What Really Downed the Hindenburg," US senator Tom Harkin, who regarded "hydrogen energy as our best hope for an environmentally safe sustainable energy future," inserted the entire *Popular Science* article into the *Congressional Record*. Bain's "incendiary-paint theory" now became the *new* orthodox explanation of the *Hindenburg* disaster.

Nevertheless, the theory attracted withering criticism. One of Bain's arguments was that the *Hindenburg* fire did not *look* like a hydrogen fire. A hydrogen flame, he said, "is almost invisible under daylight conditions." But according to eyewitness and photographic evidence, the *Hindenburg* "burnt with a very bright flame." Thus, the *Hindenburg* fire was not primarily a hydrogen fire.

Alexander J. Dessler, a physicist and himself a former NASA scientist, pointed out in a technical paper that, obviously, *both* the outer fabric *and* the hydrogen were burning, "with the hydrogen burning first and being unnoticed until it sets the fabric on fire. Then the fabric would emit light because it was burning. There would also be bright visible light produced because the fabric and the wires and girders within the fire acted like a mantle."

A mantle is a mesh cover placed over a gas jet, such as in a Coleman lantern or gas lamp, that causes it to give off a bright light when heated. A hydrogen fire made visible by a gas mantle would indeed produce a bright and noticeable flame, said Dessler. He noted further that the German zeppelins shot down during World War I had burned with a flame that could be seen for miles around, day or night, although none of them was covered with paint whose composition approximated that of the *Hindenburg*.

Dessler rejected Bain's claim that the *Hindenburg* "was literally painted with rocket fuel." "The *Hindenburg* paint differs from solid rocket fuel in an important respect. It does not contain a source of oxygen to sustain combustion." The fuel in the Space Shuttle's solid rocket boosters was made up of 69.6 percent ammonium perchlorate, a substance that provided the mixture with an oxygen source. The *Hindenburg*'s paint included no such ingredient, and thus the ship was not in fact "literally painted with rocket fuel."

But it was Norman Peake, a chemical engineer and airship enthusiast, who gave the kiss of death to the incendiary-paint theory:

Hydrogen burns through the ship while adjacent fabric remains untouched.

he observed that in photographs, "at the point where hydrogen is blazing out of the nose-vents, the surrounding fabric is intact."

For these and other reasons, Addison Bain's flammable-paint theory was preposterous as an explanation of what had happened to the *Hindenburg*. But despite its flaws, the theory had a remarkable staying power among those who considered themselves technically savvy, au courant, and "in the know."

At this point, then, there were no less than four separate and competing explanations of what had caused the *Hindenburg* fire: electrostatic ignition through grounding, St. Elmo's Fire, sabotage, and incendiary paint. What better way to decide among them than scientifically, by means of experiment? And so, for a

few years, the examination and testing of *Hindenburg* theories became a cottage industry among scientists and, as often, television
producers in search of stunning visual effects for their *Hindenburg* documentaries, reenactments, and docudramas.

In 2006 the TV series *Mythbusters* filmed a series of experimental reenactments that were at least *intended* to be scientific. The
show had built three *Hindenburg* replicas: each model was about
ten feet long, and all of them incorporated, albeit primitively, the
two classic structural elements of the zeppelin: horizontal girders
and ring sections.

The tests were far from realistic and were conducted indoors.
Worst of all, they suffered from the almost fatal limitation that
the show's presenters seemed to have taken no account of scaling
effects and made no attempt to extrapolate burn rates from their
ten-foot-long mini-models to the full-size, 804-foot-long *Hindenburg*. This was not so much science as it was show business.

The mock-ups were designed to test the incendiary-paint theory, and for that reason the first of them had been covered with
Hindenburg-like paint but had not been inflated with hydrogen.
When ignited by a blowtorch, the replica took two minutes and
six seconds to burn from end to end.

The second model was identical to the first except that the
experimenters introduced hydrogen gas into the craft—but they
did so only *after* the craft had been set afire, another lapse from
realism. Still, once it was engulfed in flames, the replica looked
eerily similar to the burning of the real *Hindenburg*. And with the
assistance of hydrogen as fuel, the craft burned twice as fast as
the empty model had, being fully consumed in fifty-nine seconds.

"To say that the hydrogen played no significant role is idiotic,"
said host Adam Savage.

It was a further shortcoming of the *Mythbusters* reenactments
that they left the possibility of sabotage out of account. But six

years later, in 2012, a British documentary filmmaker put even that theory to experimental test. At a large field belonging to the Southwest Research Institute outside of San Antonio, Texas, a film crew built three more *Hindenburg*s. These were truly immense replicas, more than 80 feet long, fully one-tenth the size of the original, and each was filled with 8,000 cubic feet of hydrogen gas. Moreover, the models reproduced many of the internal structural elements of the zeppelin—not only rings and girders but individual gas cells, ventilation shafts, and open areas at the keel that approximated the keel walkway. Covered in a bright, silvery Mylar film and set against the open sky, it was as if the *Hindenburg* had risen from the dead.

To represent the sabotage theory, the experimenters placed a miniature bomb between two of the first model's gas cells just forward of the rear fin, in a location that corresponded to where Helmut Lau had seen the fire erupt. When the film crew detonated the bomb, the model caught fire exactly where the *Hindenburg* itself had, a mushroom of flame arose above the rear of the craft, and then the whole ship started sinking to the ground in a way that precisely mimicked the fall of the original. A side-by-side video replay of the two burning craft showed the unmistakable resemblance between the two. Taken by itself, the reenactment actually *supported* the sabotage hypothesis.

The second model illustrated the theory that an electrostatic discharge had ignited free hydrogen. Ignition of the hydrogen was simulated by an induced electrical spark, but the resulting fire burned slowly, and the flaming replica did not really look like the *Hindenburg* of the newsreels.

The third and final of the film crew's three *Hindenburg*s tested the theory that St. Elmo's Fire had ignited leaking hydrogen at the top of the craft. The crew simulated that phenomenon by applying an external electrical field to the top of the model. The free hydrogen there did in fact burst into flame and then traveled down inside the ship, igniting the cells. This reenactment also looked similar to the original disaster footage.

So the sabotage reenactment looked the best of all, St. Elmo's Fire looked second-best, and the electrostatic discharge theory came in last. Nevertheless, the filmmakers decided that sabotage had to be thrown out because it could not explain the tail-heaviness of the craft during its approach to Lakehurst. Still, they did not then opt for St. Elmo's Fire as the cause but instead chose the *worst*-looking replay, the electrostatic discharge theory, on the ground that it best explained the origin and progression of the flame front. "I think the most likely mechanism for providing the spark is electrostatic," said principal filmmaker Jem Stansfield. "That starts at the top. Then the flames, from our experiments, would have probably tracked down to the center with an explosive mixture of gas that gave the *whoof* when it got to the bottom. Then the fire went up."

But all that would be true even if it had been St. Elmo's Fire, not an electrostatic discharge, that had ignited the hydrogen-oxygen mixture. Since the fire had started at the top of the craft in either scenario, there was really no basis in these experiments for choosing between the two competing explanations.

What, then, caused the *Hindenburg* disaster?

St. Elmo's Fire is a form of luminous electrical display that appears in the presence of a strong electrical field in the atmosphere. It collects around sharp points because electrical fields are more concentrated in areas of high curvature, making discharges more intense at the ends of pointed objects, and for that reason it has even been seen at the tips of cattle horns. Its luminosity is weak, however, so that the phenomenon is most often visible at nighttime.

Its bright blue or violet glow, an artifact of ionized air molecules, is also known as "brush discharge," and the two names were used interchangeably in the US Commerce Department report. It was puzzling that the investigators even considered St. Elmo's Fire at all, since the report acknowledged that "no witness testified that a visible indication of it was present." There were two possible explanations for this: one, there had been no St. Elmo's Fire to begin with, and two, "darkness had not yet fallen at the time of the accident," thus preventing St. Elmo's Fire from being seen.

As it was, St. Elmo's Fire had not been reported around the *Hindenburg* until 1962, a full quarter-century after the event, when American airship historian Douglas H. Robinson had a chance encounter with "Professor Mark Heald of Princeton, NJ, who undoubtedly saw St. Elmo's Fire flickering along the airship's back a full minute before the fire broke out."

As Robinson told the story in his book *Giants in the Sky* (1973), "Standing outside the main gate of the naval air station, he [Heald] watched as the Zeppelin approached the mast and dropped her lines. A minute thereafter, by Mr. Heald's estimate, he first noticed a dim 'blue flame' flickering along the backbone girder. There was time for him to remark to his wife 'Oh, heavens, the thing's afire.'"

But what to make of this twenty-five-year-old after-the-fact testimony was anybody's guess.

Matters were clearer concerning static electricity. Static electricity is an accumulation of an electrical charge on or in an object, and it is commonly produced by friction. The classic example, going back to antiquity, is produced by rubbing a piece of amber with a cloth. When the two objects are separated, they attract each other because they possess different charges, one positive, the other negative. Since the potential difference remains until it is removed through discharge, the electrical potential is "static," as opposed to electrical *currents,* which are dynamic. The *neutralization* of the charge differential between the two objects, however, creates a current between them, and this often occurs by means of a spark or flash of light, as, for example, in lightning.

As the *Hindenburg* flew toward its destination, it picked up the electrical charge of the surrounding atmosphere. In addition, the friction of its passage through the air added to the charge upon it, as did the further effect of the rainwater falling on the hull.

On May 6, 1937, the day of the fire, the air in the vicinity of Lakehurst was so full of electrical activity due to thunderstorms that the rubber factories in the state of New Jersey had been closed to protect workers, since static sparks were known to ignite rubber compounds. Thunderstorms had been present in the area that afternoon, and during the *Hindenburg*'s approach Commander Rosendahl had memorably reported a "thunderstorm over station."

When the ship approached the landing mast, then, it was a highly charged, "hot" object, meaning that there was a significant difference in electrostatic potential between the ship and the ground. At that point the *Hindenburg* was also tail-heavy, the most probable cause being that it had lost and was still losing hydrogen gas from one or more of the rear cells. That explains the repeated releases of water ballast from the rear, and it also explains the repeated valving of hydrogen from the forward cells, both of which made the craft heavier at the bow. And finally, it explains Captain Sammt's order of "*Six men forward!*" to further increase the weight at the nose.

While all this was going on, R. H. Ward, the ground crew member in charge of the port bow landing party, saw fabric fluttering high up at the rear of the ship, as if hydrogen gas was escaping from the ship's interior.

The ship's manila landing lines had been dropped shortly before Ward's observation, and dust flew up from the ropes as they struck the ground, showing that they were dry when they hit. The ropes were not, at that point, good conductors of electricity, but as the light rain continued to fall, their electrical conductivity increased and they progressively grounded the ship.

The landing lines were anchored to the *Hindenburg*'s metal framework, and the framework became grounded faster than its outer covering, which was less conductive than the metal. In addition, the covering was electrically isolated from the framework by wooden dowels, which the builders had installed to prevent chafing.

A gap existed where the fabric was fluttering, and a spark jumping across it would have equalized the electrical potential difference between the metal framework and the *Hindenburg*'s wet outer fabric.

The simultaneous conjunction of all of these conditions—the fluttering of the outer fabric, the leaking hydrogen, its mixing with the surrounding oxygen, the electrical differential between the outer covering and the inner metal framework, a sudden electrical discharge by means of a spark between the two—this concurrence of events was sufficient to ignite the streaming oxygen-hydrogen mixture. It caused, first, an explosion or flame that burned its way down toward the center of the ship, where it caused one gas cell to burst into flames, which in its turn led to the burning of another and then another, in a steady chain reaction toward the front of the craft, a progression that led, finally, to the destruction of the *Hindenburg*.

But the surprise was that there was nothing new about any of this. It was exactly this scenario that the Germans had presented as their explanation of the disaster in 1937. There was, in other words, no "mystery," and there never had been.

Chapter 13

THE PATHOLOGY
OF DELIRIUM

With its flames extinguished and its skeleton charred, bent, broken up, and strewn across more than 800 linear feet of ground, the *Hindenburg* as a historical artifact was still not quite dead and gone. Practically endless sections of rings and girders, as well as many of the ship's smaller parts and fittings— miscellaneous items such as fuel tanks, ladders, landing ropes, scraps of flooring, engines, gas cell netting, the nose cone and its mooring apparatus—these and other objects emerged from the crash relatively intact. Souvenir hunters carried away metal fragments, as well as pieces of mail, portions of unburned fabric, serving plates, drinking cups, cutlery, and other odds and ends, many of which, in time, would find their way into museums, where they still reside. Other bits and pieces became prized as relics of a momentous and historic event and were bought up by private collectors; every so often, one or more of them is offered for sale on eBay.

In June 1937, after the accident investigators had finished their physical examination of the *Hindenburg*'s carcass, the wreckage was sold to a salvage company in Perth Amboy, New Jersey. Some of the ship's duralumin had originally come from a previous airship wreck, that of the doomed British "socialist ship," the R 101, and now, in its turn, the mortal remains of the *Hindenburg* would be cut up, carted away, melted down, and shipped back to Germany, where the recovered metal would take on a new life, not in another zeppelin, but rather in some of the Reich's new military aircraft.

But the end of the *Hindenburg* was not yet the end of the zeppelin, as has often been claimed, nor was it the end of the zeppelin era. In June 1936, during the *Hindenburg*'s first year of operation, the Zeppelin Company in Friedrichshafen was already laying down the keel of the *Hindenburg*'s successor—in fact, its twin sister, with the same dimensions and gas capacity. The new ship would be the second coming of the *Graf Zeppelin*: the LZ 130 *Graf Zeppelin II*. Since the LZ 130 had been designed, and largely constructed, before the disaster at Lakehurst, the craft had been intended for hydrogen. But the *Hindenburg* fire changed that plan, for it was (temporarily) inconceivable to risk yet another hydrogen-filled airship catastrophe. And so zeppelinites from Hugo Eckener on down to the lowest-ranking crew member now agreed that the new ship would have to be inflated with safe, inert, and non-inflammable helium—which had been the plan for the *Hindenburg* to begin with.

But just as before, the problem was in actually getting hold of the stuff. The American Helium Control Act of 1927 was still in force, but Eckener once again hoped that it would be possible to amend the act or to obtain an exception, a waiver, or the equivalent. And so, after the Commerce Department's hearings ended at

Lakehurst, he traveled yet once more to Washington for the purpose of extracting some kind of helium concession from the US government. On May 27, 1937, with visions of the *Hindenburg* fire still fresh in their heads, the House Committee on Military Affairs listened to Eckener's testimony. "The deplorable disaster which befell the *Hindenburg* has suddenly accentuated a question which has occupied us for many years past," Eckener began. "Namely, the one of using non-inflammable gas in the zeppelins."

Committee members were generally sympathetic as Eckener proceeded to make his case, but there were plenty of bureaucratic hurdles to be overcome before he could actually lay hands on any helium and have it transported across the Atlantic. Nevertheless, he was ultimately successful in gaining a commitment from the American government to supply helium to Germany, on the condition that the Zeppelin Company would not use the substance for military purposes.

The company agreed to this and immediately began to modify the LZ 130 *Graf Zeppelin II* for the new non-inflammable lifting gas. There was a price to be paid for the changeover, however, over and above the huge expense it would entail to retrofit the craft: operationally, helium would provide about 7 percent less lift than hydrogen, which meant that the new ship would be able to carry fewer paying passengers than the number for which it had originally been designed. That load factor reduction would severely cut into the potential profits of what was already a money-losing proposition. But having no other option, the company ordered its workers to rip out the ship's existing passenger cabins, which had been designed for seventy-two people, and replace them with accommodations for forty.

After it had been modified, Charles Lindbergh visited Friedrichshafen and viewed the *Graf Zeppelin II* in its hangar. "I felt depressed looking at her," he later wrote. "This airship represented the result of all the years of development of lighter-than-air. She seemed to me like a last member of a once proud and influential family. For I can see no future for the airship. . . . Only forty passengers in such a huge machine? A plane a fraction of the size could carry more people a greater distance, and twice as fast."

Lindbergh was clearly not under the influence of the Spell, the Delirium, the Rapture, as the directors of the Zeppelin Company, a body of men who were not yet prepared to abandon the airship, continued to be. In fact, in order to achieve profitability the company was now planning an even larger craft than the *Graf Zeppelin II,* a truly colossal super-monster that could carry 100 passengers. It was the as-yet-unnamed LZ 131, an airborne specimen with the staggering length of 869 feet, more than 60 feet longer than the *Hindenburg,* to be filled with a record eight million cubic feet of helium.

All of these plans and preparations were essentially in place by the end of 1937. With a spanking new airship in the hangar and another even bigger one on the drawing boards, the zeppelin was poised to surprise everybody and stage a world comeback.

But then, on March 12, 1938, Germany annexed Austria—the *Anschluss.* The Americans saw this event as a further prelude to war with Germany, a realization that prompted Secretary of the Interior Harold L. Ickes to refuse to make any helium available to a nation that was a potential enemy. One might think that now even Germany's more hopped-up and blinkered airship fanatics might have seen the writing on the wall and renounced, disowned, and done away with their wretched zeppelins forevermore. But no. Hugo Eckener trekked to the States for what seemed like the millionth time to make a last-ditch plea for helium. He failed.

Still, the show apparently had to go on, no matter what the costs or risks. And so, having just converted it to helium, the tireless zeppelin construction crews now reconverted the LZ 130 back to the flammable and explosive hydrogen. Soon, the apparently unstoppable Eckener had dreams of taking the craft on a luxury cruise to Greenland and even making a round-trip demonstration flight to the United States. It was as if the *Hindenburg* disaster had never happened.

Beginning in August 1938, the *Graf Zeppelin II* was inflated with hydrogen, a process that took about a month, after which it would be ready to go—provided that an able and willing crew could be rounded up to fly it. Harold G. Dick, an American employee of the Goodyear Tire and Rubber Company stationed at Friedrichshafen, noted considerable uneasiness among prospective crew members about operating another hydrogen-filled zeppelin. "They had decidedly mixed feelings about flying with hydrogen," he recalled. "Every one of them had lost friends in the *Hindenburg* fire. Moreover, the very fact that a variety of precautions considered necessary were being taken—a parachute aboard for each man, a cruising altitude higher than formerly, altered landing techniques—did not add to their peace of mind." Rank-and-file workers apparently were no longer as spellbound by the airship as were Eckener and the directors of the Zeppelin Company.

Nevertheless, a crew was enlisted, and on September 14, as scheduled, Hugo Eckener baptized the ship with a bottle of liquid air and then, with himself in command, immediately launched the craft on its maiden flight. With seventy-four souls on board, the LZ 130 flew to Munich, Augsburg, and Ulm before returning to Friedrichshafen.

Two weeks later, on September 28, the ship passed a milestone when it went off on a proving flight under the command of Captain

Albert Sammt, the very same man who had been first officer on the final flight of the *Hindenburg* and who had suffered extensive burns in the disaster. As gung-ho about zeppelins as ever, Sammt had no qualms about flying the hydrogen-filled craft even through thunderstorms, which he did repeatedly for test purposes. In fact, with one exception, Sammt would command every subsequent flight of the ship, including a spy run over the coast of England and a propaganda flight over the Sudetenland (another recent Nazi acquisition), during which crew members dropped swastika flags and leaflets and blasted patriotic music to the adoring masses below, just as in the good old days of the Plebiscite flights made by the original *Graf* and the *Hindenburg* in 1936.

All this nonsense did, however, finally come to an end. In early 1940, Reich air minister Hermann Goering, never a fan of the airship, much less awestruck by it, decided that these "gasbags," as he called them, were functionally obsolete and that their metal components could be better utilized in bombers. There were, at that time, just two German zeppelins in existence: the old and the new *Graf Zeppelins*, a mismatched set, each in its own separate hangar at Frankfurt. Over the course of about a month, a Luftwaffe construction battalion dismantled them both.

"The age of the airship" ended on May 6, 1940, exactly three years to the day after the *Hindenburg* fire, when, on Goering's orders, the two remaining zeppelin hangars at Frankfurt were blown to bits. At that point, finally, the airship really was dead and gone.

But not the Delirium.

Our four pathological technologies—the zeppelin, Project Plowshare, the Superconducting Supercollider, and exotic interstellar propulsion systems and vehicles—are united by the fact that they were all shining examples of an ancient literary, aesthetic, and philosophical category—the sublime. As a designation, the sublime is most generally understood to denote the quality of greatness, an exaltation so grand, high, and mighty as to be virtually ineffable, almost beyond comprehension. Such greatness encompasses the physically huge, the extremely powerful, and the aesthetically magnificent—objects, processes, or events so imposing and astounding that they overwhelm the observer and override the mind, inducing a sort of mental intoxication, frenzy, or paralysis in the beholder.

Canonically, the sublime was synonymous with awe-inspiring natural phenomena: the Grand Canyon, Niagara Falls, the Matterhorn, volcanic eruptions, violent storms. But even in the ancient world, man-made objects such as the Seven Wonders could also be classified as sublime: the Pyramids at Giza or the Colossus of Rhodes, for example. And so could more recent human fabrications, such as the Hoover Dam or an atomic bomb explosion.

Classical theoreticians of the sublime—Longinus, Edmund Burke, Immanuel Kant—formulated sets of defining characteristics of the genre. For Burke, terror was an inescapable part of the sublime experience, so long as the observer beheld the object from a vantage point of relative safety: precipitous cliffs, threatening clouds, waterfalls, thunderstorms. Kant further distinguished between the mathematical sublime, which was incomparably and absolutely great, such as the starry heavens above, and the dynamic sublime, which involved the action of a powerful force that produced dramatic atmospheric effects, such as a hurricane or a storm at sea.

Sublime objects, furthermore, exist on a heroic scale: they are not merely big, but *big*, exciting, dangerous. Confronting them requires a break in the scale of one's ordinary perception of the world. They are unexpected, extraordinary, outside the normal course and scale of events, and these attributes combine to produce a derangement of the mind and the senses. The beholder is literally lost for words . . . and thoughts.

More recently, historian David E. Nye has introduced the concepts of "spectacular technologies" and "the technological sublime," objects such as the Saturn V space vehicle and the St. Louis Gateway Arch. A subdivision of the technological sublime is the geometrical sublime, a type of object that historically "was static and appeared to dominate nature through elegant design and sheer bulk. It found expression first in bridges and soon afterward in skyscrapers." And perhaps, too, even though they were not inherently "static," in zeppelins.

Although technologically sublime artifacts are products of human reason, they nevertheless evoke "one of the most powerful human emotions," to wit, "an essentially religious feeling, aroused by the confrontation with impressive objects." Thus, even the technologically sublime can, and often does, give rise to a temporary effacement of reason, a suspension of skepticism, a shutdown of the critical faculties.

A fundamental difference between the naturally sublime, such as Meteor Crater or the Rocky Mountains, and the technologically sublime, such as a Space Shuttle launch or the Golden Gate Bridge, is that the experience of the naturally sublime typically requires no action on the part of the observer. You simply stare at the Grand Canyon in awed amazement. You can hike into and out of the canyon, fly over it, photograph it, throw rocks into it, but there is very little that you can do to change it.

Technologically sublime objects, by contrast, are deliberate, volitional human creations and are therefore at least potentially under human control. They can be started and stopped, augmented or reduced in size, scope, and so on. Their existence through time and in the future is not guaranteed, nor in many cases either probable or desirable. However, the more sublime the object, the greater the likelihood that the flood of emotional effects it produces will be capable of influencing, even overwhelming, all rational thought regarding it. This is all the more true when the object in question mimicks or contains aspects of exceptionally powerful natural processes, such as the similarity between atomic bomb blasts and the energy releases of volcanic eruptions. In both cases, the processes are so powerful that they literally reshape the face of the earth. Thus the emotional appeal of Project Plowshare despite all of the obvious risks attendant upon nuclear-powered "planetary engineering."

The Superconducting Supercollider, likewise, emulated in its own way some of nature's primordial acts of creation: particle accelerators, themselves geometrically sublime in their immensity and complexity of design, as well as dynamically sublime in the motive power generated by their endless miles of tunnels, wires, and electromagnets, were in effect the technological equivalents of the primal universe. One of an accelerator's intended purposes was to reproduce some of the earliest moments of creation and the range of fundamental particles that existed at the time. Their architects and operators were attempting to fathom the most fundamental and deepest secrets of nature, a task that they viewed as a stirring, semireligious, almost sacred duty or calling. Hence the attachment of high-energy physicists to their immense, epochal machines, irrespective of their exorbitant costs, their overutilization of the landscape, and their scant practical usefulness to ordinary human beings.

Interstellar propulsion systems and vehicles, for their part, were technological objects on the grandest possible scale, with spaceships the size of small worlds to be propelled across immense distances at velocities that were a substantial fraction of the speed of light (or even far faster than light). These were halo machines that verged upon the infinite. They were geometrically sublime in size and dynamically sublime in action, and they held the extra added attraction that they were billed by their proponents as the means of fulfilling our one true destiny as a species. They were thus instruments of salvation, transcendental conveyances that awakened in their advocates all the evangelistic fervor of a Pentecostal revival meeting.

Terrestrial as it was by comparison to the starship, the zeppelin was nevertheless a critical nexus of emotional effects rolled into one supremely vast material object, an artifact that had no precise counterpart in the natural world other than for the fact that it flew through the air like the winged creatures of earth. In its awe-inspiring size, the zeppelin was geometrically sublime. In its ability to cross vast distances—to the point of circumnavigating the globe—it was also dynamically sublime. And above all, there was the fact that these behemoths performed the ultimate technological miracle: they defied gravity, or at least seemed to. They were ships, but they didn't plow through the ocean like whales—they actually rose up into the sky and disappeared from view, like a mass ascension of the saints into heaven.

Who could resist the resulting combination, that triple whammy of the sublime, the schizophrenic, and the absurd that was the German airship? Certainly not anyone connected with the Zeppelin Construction Company, which built the ships, or the German Zeppelin Transport Company, which flew them. The zeppelin swamped all reason, skepticism, caution, and criticism. That these airborne gas bags happened to be inflated with an explosive substance that could and did maim and kill human beings was a fact best forgotten, ignored, and papered over with the fond hope that with the very next model every possible weakness would be rectified, removed, and essentially erased from existence.

Still, for all of their considerable faults and stupidities—their huge costs, terrible risks, unintended negative consequences, and in some cases injuries and deaths—pathological technologies possess one crucial saving grace: they can be stopped.

Or better yet, never begun.

EPILOGUE

Many of those who survived the *Hindenburg* disaster returned to a reasonably normal life soon afterward, while others achieved a minor, if fleeting, measure of fame. Margaret Mather wrote an account of her experiences aboard the craft; it was published as "I Was on the Hindenburg," in *Harper's Monthly* magazine in November 1937. It was the most detailed and finely observed, albeit highly idiosyncratic, of the first-person accounts that had been written by a passenger. Despite all the tragedy, suffering, and death it caused, Mather reported that she had fond memories of the *Hindenburg*: "I could not but regret the destruction of so beautiful a thing," she wrote. "I thought of the happiness it had given to me and to many others; of the icebergs and rainbows we had flown over; I thought of how gently it had landed."

Margaret Mather completed her trip to Princeton with no further misadventures. She later returned to Rome, where she lived, and where she died at the age of ninety-one.

The other published account was Leonhard Adelt's "The Last Trip of the Hindenburg," which appeared in *Reader's Digest,* also in November 1937. It ended with the words: "The main point remains that, in the future, inflammable gas must not be used on passenger airships. England's *R 101* and Germany's *Hindenburg* are warning enough."

Leonhard and Gertrud Adelt went back to Germany and moved to Dresden. They were at home on the night of February 13, 1945, when the Allied forces firebombed the city. Leonhard, having escaped the *Hindenburg* fire with minor injuries, was badly burned in the Dresden raid and died of his burns a week later. Gertrud survived and resumed her career as a journalist.

Peter Belin, after jumping from the burning ship, finally met up with his parents on the day of the accident. They had watched in horror as the *Hindenburg* was swallowed up in flames, and they had repeatedly checked the press room blackboard to see if their son's name had appeared yet on the list of survivors. They finally gave up and returned to their car. Just as they were driving out, Peter, who had walked to the parking lot in search of his parents, saw them and flagged them down. His experience did not appear to have scarred him in any way, and he went on to a successful career in the US Navy, where he ended up as a captain.

Matilde Doehner and her two sons, Walter and Werner, remained in the hospital for ninety-two days, during which time they became momentary celebrities, much sought after by reporters. They later resumed a more or less normal life at their home in Mexico City. Werner finally ended up in a small town in a valley of the Rocky Mountains, where he still lives today, the last living *Hindenburg* survivor. For a while he gave media interviews, and he even made a trip back to Lakehurst for a reunion, but he was never really fond of talking about, or even remembering, his flight aboard the *Hindenburg*.

Marie Kleeman arrived at her daughter's home in Andover, Massachusetts, within twenty-four hours of the disaster, still in the same clothes she was wearing at the time of the crash. She even wore a Zeppelin Company souvenir pin on her coat. She eventually became a United States citizen and was buried in Andover, next to her husband Friedrich, apparently unimpressed to the end by her close escape from death.

Joseph Späh returned to his family home on Long Island, where, for an instant, he became a media star. The FBI later investigated him as possibly having been the saboteur who blew up the airship—an accusation that Späh found absurd. Both Captain Max Pruss and Chief Steward Heinrich Kubis had seen Späh as a suspicious character, but the FBI, finding no evidence that implicated Späh in any way, cleared him of wrongdoing. Späh continued on with his acrobatic career under the stage name Ben Dova and made cameo appearances in the 1976 film *Marathon Man,* starring Dustin Hoffman, and in the 1982 crime drama *Dear Mr. Wonderful.* He died in 1986, in Manassas, Virginia, where he was buried. His dog Ulla did not survive the crash.

Werner Franz, the cabin boy, testified at the US Commerce Department commission investigation and later returned to Germany. He went on to have a long life during which he gave countless interviews about his role in the disaster, was the subject of a book (*Kabinenjunge Werner Franz*), and appeared in several *Hindenburg* documentaries. In 2004 he made a trip back to Lakehurst, where he was guest of honor at the opening of a new museum at the site. He died on August 13, 2014, at the age of ninety-two.

Captain Max Pruss, the ship's commander, though severely injured in the crash, survived and recovered. He underwent several rounds of reconstructive surgery on his face, but never looked the same afterward. He always maintained that the *Hindenburg* had been sabotaged, but some later historians have suggested that this was a psychological defense intended to compensate for his culpability in the fire. The theory was that he was so intent on landing the ship quickly, since it was twelve hours late on arrival, that he took excessive risks in flying into an electrically charged area before the effects of the thunderstorms had had sufficient time to dissipate. In addition, he had ordered an "S" turn during the landing maneuvers, and some critics have claimed that the forces generated during the sharp back-and-forth turns overstressed the hull. Hugo Eckener, who himself held these actions against him, never visited Pruss in the hospital and did not speak to him until two weeks after Pruss had returned to Germany.

Charles E. Rosendahl, the Lakehurst commander, also held to the sabotage theory and wrote about it in his book, *What About the Airship?* (1938). He said of the supposed time bomb that "the device itself would not have to be much larger than an ordinary fountain pen," which, "if suspended in a gas-cell trunk, would have been hard to detect." But he offered no proof, or even a good reason to believe, that such a device had ever been aboard the *Hindenburg.*

After the dismantling of the two *Grafs* and the dynamiting of the zeppelin hangars at Frankfurt, Hugo Eckener was a man without a purpose. His entire life's work had become archaic and obsolete, and he himself had in effect become a dinosaur.

In the years that followed the *Hindenburg* catastrophe, Eckener made some halfhearted attempts to revive the commercial airship. One of them was in 1947, when he traveled to Akron, Ohio, home of the Goodyear Aircraft Corporation, for talks with Paul Litchfield, head of the company. But nothing came of this visit, and nobody in the outside world seemed to care any longer about zeppelins. Winged aircraft had long since supplanted the airship.

In 1949 Eckener published his memoirs, including his final thoughts on the destruction of the *Hindenburg,* which were substantially the same as those he'd expressed at the Commerce Department accident investigation hearings twelve years earlier. He adamantly opposed the sabotage theory and continued to think that an electrostatic discharge had ignited free hydrogen within the ship, bringing down the *Hindenburg*. He died at his home in Friedrichshafen in 1954, at the age of eighty-six.

A lthough Project Plowshare fizzled out in the 1970s, its poisonous aftereffects, unfortunately, persisted.

In 1992 it came to light that thirty years before, in 1962, a group of US Geological Survey (USGS) scientists working under contract with the Atomic Energy Commission had buried at the proposed site of Project Chariot some nuclear debris from the Sedan hydrogen bomb test of July 6, 1962. Six weeks after that explosion at the Nevada Test Site, the USGS scientists took about twenty pounds of fallout debris in the form of irradiated soil and transported it by plane up to Alaska, to Ogotoruk Creek.

On the face of it, this had been a fairly crazy operation. After all, the AEC had officially *canceled* the Chariot test and *replaced* it with the Sedan shot. Why then did scientists take some of the radioactive soil and bring it up to the Ogotoruk Creek immediately afterward? The answer was clear in light of the original and true purpose of Project Chariot, which was to discover how the fallout from a nuclear blast would percolate through a watery environment, such as in the land bordering the Panama Canal. Since there were virtually no rivers, creeks, streams, ponds, or aquifers at or near the Nevada Test Site, it was a simple case of bringing the radioactivity to the river, in effect to conduct Project Chariot more or less by proxy.

So, in August 1962, the USGS team had flown to Alaska, bringing with them samples of the contaminated soil. They had then hauled it up the banks of the Ogotoruk, where they sprinkled some of the radioactive debris on the ground and into the flowing water. Later they took water samples at various points downstream. The researchers also performed certain other tests using simulated rainwater produced by means of a gasoline-powered water pump and a garden hose. Then they collected the newly contaminated water samples and sent them back to their Denver labs for analysis.

At the end of these bizarre experiments, a bulldozer pushed about four feet of clean dirt over the top of the resulting radioactive waste pile. The team members did not mark the site, erect a fence around it, or declare it off-limits in any way. They simply left the hot materials where they were and then fled the scene. The area thereupon became essentially an abandoned and unmarked nuclear waste dump.

Thirty years later, author Dan O'Neill donated to the Point Hope village corporation some of the documents he had unearthed while researching his book about Chariot, *The Firecracker Boys*, and that was how the villagers found out about the dump. They notified the local press, whose coverage gave rise to a scandal that would ultimately involve Alaska's US senator Frank Murkowski, Alaskan governor Walter Hickel, and US secretary of defense Dick Cheney. The result of all this was that in 1993 the US government spent $6 million on a cleanup of the Chariot area. A force of fifty men dug up the still-radioactive soil and hauled it back to the Nevada Test Site—or, in Dan O'Neill's words, "back to the geniuses who put the stuff there to begin with."

When the Superconducting Supercollider project was canceled in 1993, the tunnel was about 20 percent complete, meaning that approximately ten miles of it had been hollowed out. Seventeen access shafts had been drilled as well, buildings totaling some 200,000 square feet of interior space had been constructed, and about 2,000 people were gainfully employed on the project, either at the site itself or in Dallas. After the project's cancellation, a mass exodus of scientists and other workers caused a precipitous drop in housing prices as people sold their homes and left the area forevermore. Many of the scientists would find jobs outside the field of particle physics.

The site itself, meanwhile, went the way of all entropy and neglect. Rainwater funneled down through the access shafts and collected in pools at the bottom. The aboveground buildings were ransacked, vandalized, and looted of anything valuable, and then became home to a succession of drug and alcohol parties.

In 2012, finally, the chemical company Magnablend purchased the site from Ellis County, Texas, and began renovating the office buildings. Soon the company had a workforce of ninety-seven people at the location, which they renamed the Specialty Services Complex, thereby rescuing from oblivion the letters SSC (but why?). At long last, therefore, the SSC was producing something of obvious practical value, and it was actually *making* money rather than pouring tons of it down a bottomless pit.

In January 2015, an international team of physicists announced in *Nature Physics* that they had played the grandest of all cosmic jokes on CERN and its discovery of the Higgs boson. The physicists claimed to have observed an analogue of the God particle in the superconducting materials of a low-cost desktop lab machine running at low energies. Aviad Frydman, a physicist at Israel's Bar-Ilan University and a co-director of the research, said: "Just as the CERN experiments revealed the existence of the Higgs boson in a high-energy accelerator environment, we have now revealed a Higgs boson analogue in superconductors." The CERN experiments required giga-electron volts. "The parallel phenomenon in superconductors occurs on a different scale entirely—just one-thousandth of a single electron volt," Frydman said. "What's exciting is to see how, even in these highly disparate systems, the same fundamental physics is at work."

In September 2013, the 100-Year Starship project held the second of its planned annual public symposiums at the Hyatt Regency in Houston. There was a new roster of speakers and topics, along with entertainment in the form of a $150-per-ticket performance of "Bella Gaia," a multimedia presentation of live music and dance created by the artist Kenji Williams. Researchers were proposing new interstellar propulsion systems ("A Laser Starway"), architectures for living in off-Earth environments ("3D Printed Modular Habitats"), and so on.

Otherwise, not much had changed since the symposium a year earlier. Sonny White was still working on, and explaining, his warp drive project. But now he had an intellectual ally in this strange business. David Pares, an adjunct professor of various subjects at the University of Nebraska at Omaha, was working on a warp drive of his own, in his garage, where he was trying to produce a "compression of the fabric of space," an effect that he also referred to as the phenomenon of "linear displacement." His interest in the subject stemmed from his research into Bermuda Triangle disappearances, which he also attributed to local space warps. It was all very complicated.

In March 2012, the Smithsonian Institution opened an exhibit entitled "Fire & Ice: *Hindenburg* and *Titanic*." It so happened that the year 2012 marked both the 100th anniversary of the *Titanic* sinking and the 75th anniversary of the *Hindenburg* disaster. The show, which was held at the National Postal Museum near Union Station in Washington, DC, included a selection from 301 recently discovered letters that had been rescued from the more than 17,000 pieces of mail carried on the *Hindenburg's* final flight. One of them was a charred envelope, its stamps canceled in the airship's post office on May 5, 1937, the day before the fire, that had been addressed by Herman Doehner to himself at his home in Mexico. He would not live to receive it.

The exhibit's interpretive commentary cited a number of parallels between the two vessels. Both were twentieth-century icons; both were said to be, and were, technologically advanced; both were meant for transatlantic service. They were almost equally big, physically as well as in their emotional effect upon observers. Their immense size gave to both of them a great but false sense of safety and security.

At the time they were built, they were the two largest moving objects ever constructed. And each concealed a tragic and fatal flaw.

In the case of the *Titanic,* thirty-seven seconds elapsed between the sighting of the iceberg by a lookout in the crow's nest and the moment of impact. For the *Hindenburg,* thirty-four seconds elapsed between the start of the fire and the ship's destruction. As in Robert Frost's poem, one died by fire, the other by ice. As has been said more than once, the *Hindenburg* was truly the *Titanic* of the skies.

The final irony of the *Hindenburg* disaster was that it occurred just as Hugo Eckener and company were on the brink of establishing a joint German-American airship service. The German Zeppelin Transport Company now had a counterpart in the American Zeppelin Transport Company, which was for the first time handling all of the stateside arrangements for the *Hindenburg's* transatlantic crossings, as of that final flight. There were great hopes, plans, and dreams for the future.

And then, in half a minute, all of it was lost forever.

ACKNOWLEDGMENTS

This book could not exist in anything like its present form had it not been for the invaluable assistance of Patrick Russell, who is arguably one of the world's foremost scholars of the airship *Hindenburg*. This is all the more amazing considering that Russell is not an academic historian but rather a freelance writer and editor who has spent the better part of his life researching the *Hindenburg* and writing about it, essentially as an avocation and labor of love. As a result, he probably knows more about the LZ 129 *Hindenburg,* and about the disaster at Lakehurst, than anyone alive today.

Russell is the creator of the website Faces of the Hindenburg (facesofthehindenburg.blogspot.com), which includes biographical profiles of every last passenger, crew member, and officer who flew aboard the craft on its final flight—ninety-seven people. Some of these accounts, which are profusely illustrated, are of considerable length. Further, each profile pinpoints by diagram the probable location of the person in question at the time of the crash and describes the fate of everyone involved in the tragedy.

As if all that were not enough, Russell has recently begun a second website, Projekt LZ 129 (projektlz129.blogspot.com), which incorporates information, some of it fairly technical, about the *Hindenburg* beyond what is contained in the biographical profiles just mentioned.

I have made considerable use of both websites. In addition, I am indebted to Patrick Russell for his detailed responses to my numerous requests for various arcane facts about LZ 129, and for his critical comments and corrections to my account of the last moments of the *Hindenburg* as presented in Chapter 8. It goes without saying that Russell is in no way responsible for any errors that might remain in that account, nor should it be assumed that he agrees with the views of the *Hindenburg,* or of zeppelins generally, expressed in this book.

The book's overall argument was substantially improved by the extensive critical comments and suggestions of my editor at Basic Books, T. J. Kelleher, to whom I owe many thanks. Additional thanks to my literary agent, Katinka Matson, of Brockman, Inc., for her deft assistance throughout the project, and to Quynh Do, associate editor at Basic Books, for her exceptional efficiency during production.

I am indebted to my wife, Pamela Regis, for reading and commenting on a portion of the manuscript, and for her usual sharp insights, helpful discussions, moral support, good judgment, and fine wit.

For assistance in providing documents and other source materials, advice, facts, contacts, images, and other help, I owe thanks to: Alvaro Bellon (LTA Society); Cheryl Ganz (National Postal Museum, Smithsonian Institution); An Goris (Leuven, Belgium); Dan Grossman (Airships.net); Dieter Leder (Zeppelin Study Group, Meersburg, Germany); John Provan (Kelkheim, Germany);

and Guillaume de Syon (Albright College). Thanks as well to Roger Musser and to Werner Doehner.

Finally, a special word of gratitude to Tony Reichhardt and Linda Musser Shiner of *Air & Space Smithsonian* for sending me to Houston to cover the 2012 100-Year Starship Public Symposium, an event described in Chapter 11.

SELECTED SOURCES

ZEPPELINS

Print Publications

Botting, Douglas. *Dr. Eckener's Dream Machine: The Great Zeppelin and the Dawn of Air Travel.* New York: Holt, 2001.

———. *Epic of Flight: The Giant Airships.* New York: Time/Life, 1981.

De Syon, Guillaume. *Zeppelin! Germany and the Airship, 1900–1939.* Baltimore: Johns Hopkins University Press, 2002.

Dick, Harold G., and Douglas H. Robinson. *The Golden Age of the Great Passenger Airships: Graf Zeppelin and Hindenburg.* Washington, DC: Smithsonian Institution Press, 1985.

Dooley, Sean. *The Development of Material-Adapted Structural Form.* "Appendix A-04: Aluminum, Plywood, and Rigid Airships." Thèse 2986. Lausanne: École Polytechnique Fédérale de Lausanne, 2004.

Hallion, Richard P. *Taking Flight: Inventing the Aerial Age from Antiquity Through the First World War.* Oxford: Oxford University Press, 2003.

Lehmann, Ernst A. *Zeppelin: The Story of Lighter-Than-Air-Craft.* Trans. Jay Dratler. London: Longmans, Green, 1937.

Robinson, Douglas H. *Giants in the Sky: A History of the Rigid Airship.* Seattle: University of Washington Press, 1973.

Web Sources

Airships.net: A Dirigible and Zeppelin History Site, http://www.airships
.net/.
Wikipedia, "List of Zeppelins," en.wikipedia.org/wiki/List_of_Zeppelins
(last modified March 4, 2015).

LZ 129 *HINDENBURG*

Print Publications

Archbold, Rick. *Hindenburg: An Illustrated History*. New York: Warner
Books, 1994.
Duggan, John. *LZ 129 Hindenburg: The Complete Story*. Ickenham,
UK: Zeppelin Study Group, 2002.
Robinson, Douglas H. *LZ 129 "Hindenburg."* New York: Arco, 1964.

Web Sources

Faces of the Hindenburg, "Passengers Aboard LZ 129 Hindenburg, May
3–6, 1937," October 25, 2009, facesofthehindenburg.blogspot.com.
Projekt LZ 129, "Notes on the Passenger Zeppelin, LZ 129 Hinden-
burg," projektlz129.blogspot.com.
Wikipedia, "LZ 129 *Hindenburg*," en.wikipedia.org/wiki/LZ_129
_Hindenburg.

CHAPTER 1: THE MAN IN THE SKY

Print Publications

Eckener, Hugo. *Count Zeppelin: The Man and His Work*. Trans. Leigh
Farnell. London: Massie, 1938.
Gilman, Rhoda R. "Zeppelin in Minnesota: A Study in Fact and Fa-
ble." *Minnesota History* (Fall 1965): 278–285.

Goldsmith, Margaret. *Zeppelin: A Biography*. New York: Morrow, 1931 (reprinted 1981).

Meyer, Henry Cord. *Count Zeppelin: A Psychological Portrait*. Auckland, NZ: Lighter-Than-Air-Institute, 1998.

Nielsen, Thor. *The Zeppelin Story: The Life of Hugo Eckener*. Trans. Peter Chambers. London: Allan Wingate, 1955.

Zeppelin, Count Ferdinand von. "Zeppelin in Minnesota: The Count's Own Story." Trans. Maria Bach Dunn. *Minnesota History* (Summer 1967): 265–278.

Web Sources

"Arthur Krebs, pionnier de l'aéronautique: Le dirigeable LA FRANCE de Charles Renard et Arthur Krebs,1879–1885," rbmn.free.fr /Dirigeable_LA_FRANCE_1884.HTML.

Wikipedia, "Ferdinand von Zeppelin," en.wikipedia.org/wiki/Ferdinand _von_Zeppelin.

CHAPTER 2:
THE PHILOSOPHER'S STONE OF FLIGHT

Print Publications

Cavendish, Henry. "Three Papers, Containing Experiments on Factitious Air, by the Hon. Henry Cavendish, FRS." *Philosophical Transactions (1683–1775)* (1766): 141–184.

Duval, Clément. "Pilatre de Rozier (1754–1785), Chemist and First Aeronaut." *Chymia* (1967): 99–117.

Gibbs-Smith, C. H. "Father Gusmão: The First Practical Pioneer in Aeronautics." *Journal of the Royal Society of Arts* (September 9, 1949): 822–830.

Lana [de Terzi], Francesco. *The Aerial Ship*. Aeronautical Classics No. 4. London: Aeronautical Society of Great Britain, 1910.

Web Sources

"Francesco Lana-Terzi, S.J. (1631–1687): The Father of Aeronautics," Fairfield University, Mathematics Department, www.faculty.fair field.edu/jmac/sj/scientists/lana.htm.
Wikipedia, "Bartolomeu de Gusmão," en.wikipedia.org/wiki/Bartolomeu _de_Gusmão.
Wikipedia, "Henry Cavendish," en.wikipedia.org/wiki/Henry_Cavendish.

CHAPTER 3: THE FLYING BOMB

Print Publications

Asimov, Isaac. *Asimov's Biographical Encyclopedia of Science and Technology.* 2nd rev. ed. Garden City, NY: Doubleday, 1982.
Dooley, Sean. *The Development of Material-Adapted Structural Form.* Appendix A-04: Aluminum, Plywood, and Rigid Airships. Thèse 2986. Lausanne: École Polytechnique Fédérale de Lausanne, 2004.
Hirschel, Ernst Heinrich, Horst Prem, and Gero Madelung. *Aeronautical Research in Germany: From Lilienthal Until Today.* Vol. 147. Berlin: Springer, 2004.
Robinson, Dr. Douglas H. "Review of Cvi Rotem: *David Schwarz: Tragodie Des Erfinders.*" Bloomington: Indiana University Printing Services, 1983. In *Buoyant Flight,* 2–8 (Akron, OH: Lighter-Than-Air Society, March 1984).

Web Sources

"The Construction and Testing of the [Schwarz] Airship," croatian-treasure.com/airconst.html.
"Ferdinand Graf Zeppelin" (navigable balloon, US patent 621195-A, patented March 14, 1899), google.com/patents/US621195.
Wikipedia, "David Schwartz (Aviation Inventor)," en.wikipedia.org /wiki/David_Schwarz_(aviation_inventor).

CHAPTER 4: THE DELIRIUM

Print Publications

Chollet, Capt. L. "Balloon Fabrics Made of Goldbeater's Skins." *L'Aéronautique* (August 1922): 258–262. Trans. National Advisory Committee for Aeronautics (December 1922).

Gann, Ernest K. *Ernest K. Gann's Flying Circus.* New York: Macmillan, 1974.

Steadman, Mark. "The Goldbeater, the Cow, and the Airship" (May 1, 2006). Post & Tele Museum (Copenhagen, DK).

CHAPTER 5: DEMYSTIFYING GARGANTUA

Print Publications

Dürr, Dr. Ludwig. *25 Years of Zeppelin Airship Construction.* Trans. Alastair Reid. Berlin, 1924.

Eckener, Dr. H. "Brief Instructions and Practical Hints for Piloting Zeppelin Airships" [1919]. Reprinted in Douglas H. Robinson, *LZ 129 "Hindenburg."* New York: Arco, 1964.

CHAPTER 6: A TECHNOLOGICAL ANOMALY

Print Publications

Brock, William H. *The Norton History of Chemistry.* New York: Norton, 1992.

Kelly, Kevin. *What Technology Wants.* New York: Penguin, 2010.

Shute, Nevil. *Slide Rule: An Autobiography* [1954]. Cornwall, UK: House of Stratus, 2000.

Web Sources

Wikipedia, "USS *Akron* (ZRS04)," en.wikipedia.org/wiki/USS_Akron _(ZRS-4).

CHAPTER 7:
DEATH RATTLE OF A LEVIATHAN

Print Publications

Adelt, Leonhard. "The Last Trip of the Hindenburg." *Reader's Digest* (November 1937): 69–72.

Mather, Margaret G. "I Was on the Hindenburg." *Harper's Monthly* (November 1937): 590–595.

Web Sources

"Airship Drawings by David Fowler" [technical design drawings of the *Hindenburg* LZ-129 and the *Graf Zeppelin* (II) LZ-130], www .highriskadventures.com/airships (published 2009).

Airships.net: A Dirigible and Zeppelin History Site, http://www.airships .net/.

British Pathé, "Hindenburg Disaster Real Footage, 1937," www.youtube .com/watch?v=CgWHbpMVQ1U (uploaded July 27, 2011).

"Hindenburg Footage from Aboard the Last Flight" [Joseph Späh video], www.youtube.com/watch?v=0rSzc8JxBFg (uploaded September 30, 2008).

"USS *Akron* Accident—February 22, 1932," www.youtube.com/watch ?v=P4MWd1pKvZI (published April 4, 2013).

CHAPTER 8: FROM HUBRIS TO HORROR
IN THIRTY-FOUR SECONDS

Web Sources

Faces of the Hindenburg, "Passengers Aboard LZ 129 Hindenburg, May 3–6, 1937," October 25, 2009, facesofthehindenburg.blog spot.com.

Projekt LZ 129, "Notes on the Passenger Zeppelin, LZ 129 Hindenburg," projektlz129.blogspot.com.

CHAPTER 9:
PROGRESS THROUGH H-BOMBS

Print Publications

"Caribou May Bar Alaska A-Blast." *New York Times,* June 4, 1961.

Kauffman, Scott. *Project Plowshare: The Peaceful Use of Nuclear Explosives in Cold War America.* Ithaca, NY: Cornell University Press, 2013.

Kirsch, Scott. *Proving Grounds: Project Plowshare and the Unrealized Dream of Nuclear Earthmoving.* New Brunswick, NJ: Rutgers University Press, 2005.

O'Neill, Dan. *The Firecracker Boys.* New York: St. Martin's, 1994.

Reines, Frederick. "Are There Peaceful Engineering Uses of Atomic Explosives?" *Bulletin of the Atomic Scientists* (1950): 171–172.

Rhodes, Richard. *The Making of the Atomic Bomb.* New York: Simon & Schuster, 1986.

Teller, Dr. Edward. "We're Going to Work Miracles." *Popular Mechanics* (March 1960): 97–101, 278–280.

US Public Health Service. "Iodine Inhalation Study for Project Sedan." Unpublished manuscript, May 20, 1964.

Web Sources

"Executive Summary: Plowshare Program—OSTI (Office of Scientific and Technical Information)," www.osti.gov/opennet/reports /plowshar.pdf.

CHAPTER 10: THE GODZILLA OF PHYSICS

Print Publications

Appell, David. "The Supercollider That Never Was." *Scientific American* (October 15, 2013).

Anderson, Philip. "The Case Against the SSC." *The Scientist* (June 1, 1987).

Blau, Steven K. "Particle Acceleration on a Chip." *Physics Today* (October 2013).

Breuer, John, and Peter Hommelhoff. "Laser-Based Acceleration of Nonrelativistic Electrons at a Dielectric Structure." *Physical Review Letters* (2013).

Congressional Budget Office. *Risks and Benefits of Building the Superconducting Super Collider.* Washington, DC: US Government Printing Office, 1988.

Dyson, Freeman. "Alternatives to the Superconducting Super Collider." *Physics Today* (February 1988): 77.

———. *From Eros to Gaia.* New York: Pantheon Books, 1992.

———. *Imagined Worlds.* Cambridge, MA: Harvard University Press, 1997.

———. *Edward Teller 1908–2003: A Biographical Memoir.* Washington, DC: National Academy of Sciences, 2007.

Hilts, Philip J. *Scientific Temperaments: Three Lives in Contemporary Science.* New York: Simon & Schuster, 1982.

Horgan, John. "If You Want More Higgs Hype, Don't Read This Column." *Scientific American* (July 4, 2012).

Kevles, Daniel J. "Good-bye to the SSC: On the Life and Death of the Superconducting Super Collider." *Engineering and Science* (Winter 1995): 16–25.

Lederman, Leon, with Dick Teresi. *The God Particle: If the Universe Is the Answer, What Is the Question?* New York: Houghton Mifflin, 1993.

Limon, P., et al. *Design Study for a Staged Very Large Hadron Collider.* Fermilab-TM-2149. Batavia, IL: Fermilab, 2001.

Overbye, Dennis. "Finding the Higgs Leads to More Puzzles." *New York Times,* November 4, 2013.

Peralta, E. A., et al. "Demonstration of Electron Acceleration in a Laser-Driven Dielectric Microstructure." *Nature* (September 27, 2013).

Regis, Ed. "What's the Hurry?" (review of Steven Weinberg, *Dreams of a Final Theory*). *London Review of Books* (June 24, 1993): 18–19.

Sherman, Daniel, et al. "The Higgs Mode in Disordered Superconductors Close to a Quantum Phase Transition." *Nature Physics*
(2015): 188–192.

Sterling, Bruce. "The Dead Collider." *Magazine of Fantasy and Science
Fiction* (July 1994).

Subcommittee on Contracting Practices for the Superconducting
Super Collider. *Contracting Practices for the Underground Con-
struction of the Superconducting Super Collider.* Washington, DC:
National Academies Press, 1989.

Weinberg, Steven. *Dreams of a Final Theory.* London: Hutchison Ra-
dius, 1993.

———. "The Crisis of Big Science." *New York Review of Books*
(May 10, 2012).

Wines, Michael. "House Kills the Supercollider, and Now It Might
Stay Dead." *New York Times,* October 20, 1993.

CHAPTER 11: STARDATE 90305.55

Print Publications

Alcubierre, Miguel. "The Warp Drive: Hyper-Fast Travel Within General Relativity." *Classical and Quantum Gravity* (1994): L73–L77.

Andreesen, G. B. "Trapped Antihydrogen." *Nature* (November 17, 2010).

Comins, Neil F. *The Hazards of Space Travel: A Tourist's Guide.* New
York: Villard, 2007.

Cresswell, Matthew. "How Buzz Aldrin's Communion on the Moon
Was Hushed Up." *The Guardian,* September 13, 2012.

Dyson, Freeman J. "Interstellar Transport." *Physics Today* (October
1968): 41–45.

Garber, Megan. "The Trash We've Left on the Moon." *Atlantic* (December 19, 2012).

Gingell, Tom W. "Starship Collision Warning Using Quantum Radar."
100YSS 2012 Public Symposium. Houston, September 14, 2012.

Heppenheimer, T. A. "On the Infeasibility of Interstellar Ramjets."
Journal of the British Interplanetary Society (1978): 222.

Kakaes, Konstantin. "Warp Factor." *Popular Science* (April 2013).

Overbye, Dennis. "Offering Funds, US Agency Dreams of Sending Humans to Stars." *New York Times,* August 17, 2011.

Sagan, Carl. *Cosmos.* New York: Random House, 1980.

Smith, Cameron M. "Starship Humanity." *Scientific American* (January 2013): 39–43.

Thompson, David. "Astropollution." *CoEvolution Quarterly* (Summer 1978): 34–51, 104–105.

Wolfe, Tom. *The Right Stuff.* New York: Farrar, Straus and Giroux, 1979.

Web Sources

CNN, "What an Enterprise! NASA Physicist, Artist Unveil Warp-Speed Craft Design," June 12, 2014, cnn.com/2014/06/12/tech/innovation/warp-speed-spaceship.

Marc Millis, "100 Year Starship Meeting: A Report, by Paul Gillster," January 28, 2011, centauri-dreams.org/?p=16525.

"100 Year Starship Study: Exploring the Future of Space Travel," 100yearstarshipstudy.com.

Harold "Sonny" White, "Warp Field Mechanics 101," NASA Johnson Space Center (2011), go.nasa.gov/ZPWZnM.

Wikipedia, "Alcubierre Drive," en.wikipedia.org/wiki/Alcubierre_drive.

Wikipedia, "Interstellar Travel," en.wikipedia.org/wiki/Interstellar_travel.

Wikipedia, "Mae Jemison," en.wikipedia.org/wiki/Mae_Jemison.

CHAPTER 12: THE SIX *HINDENBURGS*

Print Publications

Bain, Addison, and William D. Van Vorst. "The Hindenburg Tragedy Revisited: The Fatal Flaw Found." *International Journal of Hydrogen Energy* (1999): 399–403.

Dessler, A. J., D. E. Overs, and W. H. Appleby. "The *Hindenburg* Fire: Hydrogen or Incendiary Paint?" *Buoyant Flight* (January–February and March–April 2005).

The Hindenburg (film), directed by Robert Wise, featuring George C. Scott, Anne Bancroft, William Atherton, Roy Thinnes, Gig Young, Burgess Meredith, Charles Durning, and Richard A. Dysart. Universal Pictures, 1975 (DVD, 1998).

The Hindenburg (TV show). A&E Television Networks, 2009.

Hoehling, A. A. *Who Destroyed the Hindenburg?* Boston: Little, Brown, 1962.

Knight, R. W. "Report No. 11: The Hindenburg Accident." US Department of Commerce, 1938.

Mooney, Michael Macdonald. *The Hindenburg.* New York: Dodd, Mead, 1972.

Mythbusters (episode 70). Discovery Channel, broadcast January 10, 2007.

Schwartz, John. "The Best Science Show on Television?" *New York Times,* November 21, 2006.

Tonmitr, Kittipong, and Arkom Kaewrawang. "Saint Elmo's Fire Corona by Using HVDC, HVAC, and Tesla Coil." *Marine Engineering Frontiers* (May 2013): 19–23.

"The True Story of Hydrogen and the 'Hindenburg' Disaster," *Congressional Record,* 105th Cong., 2nd sess., vol. 144, issue 138 (Senate, October 6, 1998), S11631.

Vaeth, J. G. "What Happened to the Hindenburg?" *Weatherwise* (1990): 315–323.

What Destroyed the Hindenburg? Discovery Channel, broadcast December 16, 2012.

Web Sources

Addison Bain and Ulrich Schmidtchen, "Afterglow of a Myth: Why and How the 'Hindenburg' Burnt," www.dwv-info.de/e/publications /2000/hbe.pdf.

C-Span, "May 6, 1937: Hindenburg Disaster" (video), http://www .c-span.org/video/?305403–1/hindenburg-disasteren.wikipedia.org /wiki/Hindenburg_disaster.

A. J. Dessler, "The Hindenburg Hydrogen Fire: Fatal Flaws in the Ad-
    dison Bain Incendiary-Paint Theory," June 3, 2004, spot.colorado
    .edu/~dziadeck/zf/LZ129fire.pdf.
Federal Bureau of Investigation, "FBI Records: The Vault" (investigation
    of *Hindenburg* disaster), vault.fbi.gov/Hindenburg%20/Hindenburg
    %20Part%201%20of%204/view.
Dan Grossman, "The Hindenburg Disaster," Airships.net, airships.net
    /hindenburg/disaster.
Hoehling v. Universal City Studios, Inc., LOIS (Law Office Informa-
    tion System) Law Library, law.cornell.edu/background/amistad
    /6182d972.htm.
Projekt LZ129, "Das verstehe ich nicht . . . " ("I don't understand
    it . . . "), January 11, 2013, projektlz129.blogspot.com/2013/01
    /das-verstehe-ich-nicht.html.
Projekt LZ129, "Sechs Männer nach vorne" ("Six men forward"), Octo-
    ber 26, 2013, projektlz129.blogspot.com/2013/10/sechs-manner
    -nach-vorne.html.

CHAPTER 13:
THE PATHOLOGY OF DELIRIUM

Print Publications

Lindbergh, Charles A. *The Wartime Journals of Charles A. Lindbergh*.
    New York: Harcourt Brace, 1970.
Nye, David E. *American Technological Sublime*. Cambridge, MA:
    MIT Press, 1994.

EPILOGUE

Print Publications

Ganz, Cheryl R., and Daniel Piazza, with M. T. Sheahan. *Fire & Ice:
    Hindenburg & Titanic* (exhibit booklet). Washington, DC: Smith-
    sonian National Postal Museum, 2012.

Jayalakshmi, K. "'God Particle' Discovered in Low Cost Experiment Using Superconductors." *International Business Times,* February 21, 2015.

Sherman, Daniel et al. "The Higgs Mode in Disordered Superconductors Close to a Quantum Phase Transition." *Nature Physics* (2015): 188–192.

Web Sources

Casey Logan, "Working Toward a Warp Drive: In His Garage Lab, Omahan Aims to Bend Fabric of Space," December 26, 2014, Omaha. com, http://www.omaha.com/living/working-toward-a-warp-drive -in-his-garage-lab-omahan/article_b6489acf-5622–5419-ac18 –0c44474da9c9.html.

ILLUSTRATION SOURCES

Page v: Associated Press [*Hindenburg* over Manhattan]
Page 78: Library of Congress [LZ 4 at Echterdingen]
Page 84: Luftschiffbau Zeppelin GmbH [LZ 5 at Weilberg]
Page 89: Luftschiffbau Zeppelin GmbH [LZ 8, *Deutschland II,* at Düsseldorf]
Page 241: NASA [Lunar landfill]
Page 259: Associated Press [Bow torch]

INDEX

PEPI KHARA

Ed Regis is a longtime science writer; the author of eight books, including *What Is Life?*, *The Info Mesa*, and *Who's Got Einstein's Office?*; and the coauthor with George Church of *Regenesis: How Synthetic Biology Will Reinvent Nature and Ourselves*. Regis lives in Maryland.